Out of Eden

Out of Eden
The Surprising Consequences of Polygamy

David P. Barash

OXFORD
UNIVERSITY PRESS

OXFORD
UNIVERSITY PRESS

Oxford University Press is a department of the University of Oxford. It furthers
the University's objective of excellence in research, scholarship, and education
by publishing worldwide. Oxford is a registered trade mark of Oxford University
Press in the UK and certain other countries.

Published in the United States of America by Oxford University Press
198 Madison Avenue, New York, NY 10016, United States of America.

© Oxford University Press 2016

First Edition published in 2016

Cataloging-in-Publication data is on file at the Library of Congress
ISBN 978–0–19–027550–1

1 3 5 7 9 8 6 4 2
Printed by Sheridan, USA

Contents

1

Polygamy 101

Romeo and Juliet, Tarzan and Jane, Ozzie and Harriet, and of course, Adam and Eve—adorable, admirable, appealing, . . . and one thing more: mythic. In one form or another, the iconic image of the happy, naturally monogamous, heterosexual twosome has been with us for a long time, but it has always been a fiction. There is something downright inspiring about that First Couple sharing biblical bliss—albeit, nonsexual—in the Garden of Eden; and then, when banished, striding out hand in hand, ready to confront the world as a committed couple. But evolutionary biologists and anthropologists know that reality is otherwise. Not only are we the products of evolution rather than special creation (corrective reality 1), but human beings evolved in a regime in which mating was often polygamous; and, moreover, we carry stigmata of this circumstance with us today (corrective reality 2, and the subject of this book).

Just as the Biblical account of Adam and Eve has been superseded by the evolutionary account of how we actually came to be, the Edenic myth of mutually embraced monogamy is, right now, being replaced by yet more biological understanding of how our sexual selves actually evolved and how as a result men and women are inclined—when possible—to mate with more than one member of the opposite sex. We never actually lived in any Garden of Eden, but many of our ancestors—not that long ago—were polygamous. This fact, and the often-hidden consequences of our inherited polygamous inclinations, have had a number of unexpected and largely troublesome effects.

Only recently, however, have anthropologists, evolutionary biologists, and psychologists understood what has been going on and why, not just in our sex lives but in other seemingly disconnected ways that human beings go about being, well, human. The biological reality is that we weren't "made for monogamy," despite the preferences of straight-laced (and often hypocritical) preachers; and not for free-spirited sexual adventurism either, despite the fervent desire of those seeking to justify a chosen "swinging" lifestyle. *Out of Eden* isn't intended to be prescriptive (it won't tell you what to do); nor will it be proscriptive (announcing what *not* to do). By the same token, it doesn't advocate any particular lifestyle, whether strict marital fidelity or libidinous polyamory. I am committed to telling it like it is, letting the chips fall where they may, letting it all hang out—choose your cliché—in short, adhering rigorously to my understanding of what science tells us[1] and refraining from telling anyone, in turn, what he or she *ought* to do.

This book, in short, is science, not advocacy. In writing it, I have made use not only of my own research but also of recently published work, where appropriate. I have also not been shy about referring to studies that in some cases are as much as several decades old and that—in this era of instant online publication, information overload, and consequent short public attention span—have not received the attention they deserve. In many such cases, even though research scientists have long ago incorporated such findings into their work, their significance (even their existence) has gone largely unrecognized by the general public.

In an earlier book, *The Myth of Monogamy: Fidelity and Infidelity in Animals and People*,[2] my wife Judith Eve Lipton and I coauthored the first accessible written account of findings based on DNA fingerprinting, which showed that sexual monogamy (as opposed to "social monogamy") is exceedingly rare—in humans as well as other animals. In the book you are now reading, I build on that, incorporating research that has accumulated since 2001, clearly demonstrating the underlying prevalence of polygamy as the default setting for human intimacy. I also develop some of the wide-ranging consequences of this new understanding of human nature, presenting material that is for the most part widely acknowledged by biologists and anthropologists but is made available here, to the general public, for the first time.

It isn't just old habits that die hard; old preconceptions can be at least as stubborn. Just think of how much easier it is to change your clothes than to change your mind. This is especially true when it comes to the nature of human nature, an arena in which scientific findings often confront a reluctance to entertain ideas that differ from what we have been taught, involving matters both of self-image and wishful thinking, not to mention—as in our case—some of the most basic prior teachings of ethics, morality, and religious doctrine. But the truth has a habit of emerging.

More than two millennia ago, the ancient Greek aphorism "Know Thyself"— attributed to Socrates, among others—was inscribed at the entrance to the Delphic Temple of Apollo. The present book is a contribution toward precisely this self-knowledge. It is likely to be controversial, in no small part because it takes issue with the postmodernist contention that "human nature" (along with much of what's left of objective reality) is merely a social construct. It is particularly likely to arouse the ire of those who argue that essentially all male–female differences, except for the basic plumbing, are socially rather than biologically founded—and/ or that they result from competing "narratives" reflecting the clash of divergent power structures—and are therefore nothing but verbal constructs that say much about the interests and illusions of those employing them but nothing about what is actually true.

Although I confess to being somewhat immodest,[3] I cannot claim that the material here is the final word on human sexual inclinations, mating systems, violence, parenting, and so forth. But I can state, unblushingly, that it accurately reflects the current state of the relevant science: that is, what we know, at present, to be "actually true" about ourselves.

The best view in Warsaw, Poland, is from the top of the Palace of Science and Culture, an example of Stalinesque architecture at its worst. Why is this so? Because this building is pretty much the only place in the city from which one cannot see the Palace of Science and Culture! The moral is that it is difficult to see something when you are very close to it. Although this is useful information for anyone stuck in Warsaw who might be allergic to bad architecture, it is a genuine problem for those interested in understanding not buildings but people. It is notoriously difficult to get a good, close, objective look at our own species, our selves. Nonetheless, evolutionary biology offers an especially clarifying perspective.

It is one thing to know ourselves—more accurately, to pursue such knowledge, although the process is less like finding a Holy Grail than like pursuing the horizon because even as we cover a lot of ground, ultimate success remains elusive. It is quite another when this pursuit also mandates that en route, we give up some of our most cherished notions. In the pages to come, I will not be arguing that "biology is destiny" but quite the opposite, that we are most free from biological constraints in proportion to our understanding of those inclinations and predispositions with which evolution has endowed us.

For a long time it was unacceptable to demonstrate (or even to suggest) that there are some behavioral differences between men and women that are primarily due to their differing biology. And in many quarters, this is still the case. However, primatologist Sarah Hrdy points out that "Among feminist scholars it is now permissible to say that males and females are different, provided one also stipulates that females are more cooperative, more nurturing, more supportive—not to mention equipped with unique moral sensibilities."[4] To some extent, this is actually true: women really are more cooperative, more nurturing, and more socially supportive. As we'll see, there are perfectly good (i.e., biologically based) reasons why this is so.

In any event, knowing ourselves requires more than acquiring knowledge per se; we must also toss out false teachings and erroneous preconceptions, emulating not just the Greek intellectual hero Socrates but also that muscular paragon Hercules whose most odious task was to clean out the Augean Stables. In following our species' journey out of Eden and into the 21st century, we'll have to do some pretty serious shoveling.

Unlike the demands placed on Hercules, however, our task will require not physical strength but intellectual courage. Seeking a motto for the Age of Enlightenment, that exciting time when Europeans, in particular, were much taken with the prospect of using reason and science to penetrate the many secrets of nature, philosopher Immanuel Kant suggested *Sapere aude*, "Dare to know." I am happy to resurrect it here and now, as both a challenge and a hope applied to some of our most personal and hitherto unacknowledged inclinations.

Sapere aude, everyone!

Polygamy isn't only troublesome ethically, socially, economically, and legally: it's also complicated in its own right. It comes in two forms: *polygyny*, in which one man maintains a "harem" of wives with whom he mates and typically produces children; and also its mirror image, *polyandry*, whereby one woman mates with more than one man. Polyandry is exceedingly rare, at least as a formally constituted mating system complete with recognized marriages in which one wife has multiple acknowledged husbands. (It is best known from certain tribal groups in Tibet, the Marquesa Islands, and some regions in the Brazilian rainforest and pretty much nowhere else.) Polygyny, on the other hand, is much more widely distributed, both geographically and throughout recorded history; more important, its imprint is detectable in some of our deepest inclinations.

The reason "polygamy" is often used as a synonym for "polygyny" is probably because polyandry is so rare that polygamy and polygyny are almost the same thing. But they aren't. In fact, although polygyny is well known—not only because it is very common but because it represents a pattern that is often culturally encouraged—the reality is that polyandry is also common: not as a formally constituted marital system but rather simply because women, like men, are also prone to having multiple sexual partners. It's just that in the context of patriarchal social traditions (which is to say, most systems!), polyandry is typically hidden, whereas polygyny is more overt.

Both mating variants, polygyny and polyandry, are fascinating as well as fundamental to being human. Taken as a whole, polygamy—that is, polygyny plus polyandry—is politically fraught, emotionally disruptive, economically confusing, illegal (at least in modern Western societies, including the United States): and yet it is something to which all of us, like it or not, are biologically predisposed.

Polygyny in particular is surrounded by myths and misunderstandings, even though its roots go deep into our evolutionary history; and its impact can be seen not only in our anatomy and physiology but also in many of our behavioral predispositions (including sexual preferences as well as inclinations for violence and many other things) plus certain seemingly unbiological traits, such as parenting and creativity, and maybe even homosexuality and monotheism.

The polygamous nature of *Homo sapiens*, although clearly demarcated by biology, is likely to be controversial and "politically incorrect." But that doesn't make it any less valid, or render it any less important as something to acknowledge and understand. *Sapere aude*, remember? If a Martian zoologist were to visit the Earth, there is no question that he or she—or it—would conclude right away that *Homo sapiens* is polygamously inclined. Evolutionary biologists are well aware of these facts; but so far, the general public has been kept in the dark, perhaps because the situation is so disconcerting. Time to remove the fig leaf. Here, for starters, is a summary of the major evidence for polygyny (male-based harem formation):

> Finding 1: In all polygynous species, males are physically larger than females. Basically, this is because polygyny produces a situation in which males compete with each other for access to females; and in the biological arena, such competition typically occurs via direct physical confrontations in which the larger and stronger nearly always wins. If the species is strictly polygynous—that is, if polygyny is the only mating system (such as in elk, or gorillas)—then a small number of males get to father many offspring, whereas most males are reproductive losers, ending up as unmated, frustrated, and resentful bachelors. The greater the "degree of polygyny" (essentially, the larger the average harem size), the more bachelors.

This predictable imbalance between the numbers of mated and unmated males is an unavoidable result of the fact that there are typically equal numbers of males and females in every sexually reproducing species. Consider a species of wild deer whose degree of polygyny is quite modest and in which the average harem master collects, defends, and mates with four females; in such a case, although pretty much every female ends up becoming a mother, there are on average three unsuccessful bachelor males for every harem holder. Compare this with an imaginary species with a much higher degree of polygyny: say, 100 females to a harem. In this case, there are 99 bachelor males for every harem keeper. The inevitable result is that whereas each harem master has about 100 offspring every year, there are also 99 excluded males whose reproductive performance is zero. By contrast, because every female is likely to be a member of one harem or another, essentially all females get to reproduce, as a member of some harem master's coterie.

This disparity between the patterning of male and female reproductive success within polygynous species turns out to be very important. Another more technical way of saying this is that under polygyny, the "variance" in male reproductive success is high, whereas the variance in female reproductive success is low.[5] Consider elk, for example, and the realistic assumption that a harem-holding bull may father 20 offspring (one with each of "his" females), leaving 19 unmated and unreproductive bachelor bulls. By contrast, every cow elk is likely to be impregnated by the harem master. Given this situation, if you are a bull elk, there is a huge payoff to being the harem master and a huge cost to being one of the subordinate, excluded bachelors. That payoff comes directly in reproductive success ("fitness" in evolution speak), something that is achieved big time by harem keepers—and the greater the harem, the larger the fitness payoff. By the same token, bachelors are evolutionary failures whose combined fitness is as low as the harem keeper's is high.

This outcome is very consequential: natural selection favors bulls who are physically imposing and therefore successful in bull–bull confrontations because they are the ones whose genes are projected into the future, whereas there is no comparably unbalanced payoff for cows. For bulls, the situation is close to winner take all (all the females available in any given harem, along with all the reproductive success). For cows, everyone is, to some degree, a winner—although no one wins inordinately. One interesting result, incidentally, is that because they are largely freed from the tribulations of same-sex competition, females often get a curious benefit: they are more likely than males to be at the ecological optimum when it comes to body size. Males, on the other hand, because they are constrained by the rigors of sexual competition, are more likely to be too big for their own good.[6]

As anthropologist/primatologist Sarah Hrdy points out,

> the virtues of large size are not limitless. Even though most bulls don't live in china shops, there are, nevertheless, costs attached to that much bulk. Limitations to male size include availability of food and the restrictions of gravity. Orangutans are among the most arboreal of apes, yet a fully grown male (weighing up to 165 pounds) may become so large that the forest canopy

no longer supports his weight and he is forced to travel long distances by walking along the tangled, leech-infested floor. Larger than the female by 25 to 50%, the male orangutan is confined by his foraging needs to a nearly solitary existence. Slowly, persistently, and endlessly, the shaggy, phlegmatic red titan consumes vast quantities of unripe fruits and mature leaves, the junk foods left by more discriminating females. The female orang, because she is smaller, can afford to be a picky eater, selecting the nutritious shoots of new leaves and the ripest fruits.[7]

Why then are males pushed into being supersized? This is easily answered once we ask who is likely to win when it comes to male–male competition in a polygynous species. The victors (those favored by natural selection) are nearly always those that are larger and stronger, which in turn selects for larger size in the more competitive, harem-keeping sex. As a result, when any species shows a consistent pattern of males larger and stronger than females, it's a good bet that polygyny is involved. Greater male size isn't by itself proof of polygyny, but it points in that direction.

Furthermore, there is a clear correlation between the degree of polygyny—essentially, harem size—and the degree of male–female size difference (the technical term for such difference is "sexual dimorphism"). Among monogamous species, males and females are pretty much physically matched because neither sex is especially subject to heavy competition when it comes to obtaining mates. Gibbons, for example, are the only species of great apes whose mating system is close to monogamous, and the body sizes of male and female gibbons is very close to equal.

Among mildly polygynous species, males are somewhat larger, and among highly polygynous ones, the male–female difference is very great. When it comes to *Homo sapiens*, men are somewhat larger than women (roughly by a factor of 20%), well within the range of other, moderately polygynous species. Note, as well, that in this and other cases, we are dealing with a statistical generalization, which is not invalidated by the fact that *some* men are indeed smaller than *some* women. The fact remains that by and large, men are larger and physically stronger than women. Not coincidentally, by the way, women are "stronger" in that they live longer, something probably due in large part to the rigors of male–male competition . . . of which there will be more later (see chapter 2, "Violence").

There is debate among physical anthropologists, paleontologists, and like-minded specialists over the extent of sexual dimorphism among our evolutionary ancestors, particularly the Australopithecines of whom "Lucy" is the best known. The consensus is that such dimorphism was quite high, strongly suggesting that polygyny was widespread, perhaps even as well-developed and extreme as currently found among gorillas.[8] On the other hand, there is no evidence of sexual dimorphism in canine teeth within the evolutionary line leading to and including modern humans (known technically as the "hominins"). Canine teeth are useful not only for catching prey but also for threatening and actually acting out violent competition of the sort associated with mating success.

Hence, the fact that ancestral hominins lacked canine-tooth dimorphism is a vote suggesting that our ancestors—previously polygynous, like most mammals—were actively evolving toward some other system, perhaps a multimale, multifemale band organization like modern bonobos or chimpanzees, or even something approaching monogamy,[9] as with modern gibbons. Others maintain that canine tooth dimorphism isn't a useful indicator of mating system in hominins because among these primarily bipedal creatures, male–male competition would have occurred not via teeth (after all, the hominin jaw is pronouncedly recessed, rather than protruding like the muzzle of a cat, dog, or even baboon or gorilla) but with use of the arms and hands, and quite possibly employing crude weapons as well.

Men are about 20% heavier than women. It is noteworthy, in any event, that the ratio of male to female overall body muscle mass in modern human beings is 27.4 to 17.9 kg,[10] yielding a sexual dimorphism muscle mass ratio of 1.53, as contrasted with a simple body mass ratio of 1.2. This enhanced difference when it comes to muscle mass could reflect either the effects of sexual selection in a polygynous mating system or selection for male success in hunting large prey items, or for defense against terrestrial predators as well as against marauding human groups, or all of the preceding, possibly in different combinations in different local populations. Another way of looking at this: when fat is discounted (more abundant in women because of the advantage it confers during pregnancy and lactation), men are 40% heavier, with 60% more muscle mass and a whopping 80% more muscle in their arms.[11]

On the other hand, it isn't clear why antipredator or antimarauder protection should be more selected in polygynous than in monogamous mateships, and yet sexual dimorphism is consistently more highly developed in the former cases than in the latter.

Primatologist Alan Dixson, concluding a detailed review of *Sexual Selection and the Origins of Human Mating Systems* (the title of his impressive book), concluded that

> It is likely that *Homo sapiens* evolved from a primarily polygynous non-human primate precursor, and that the earliest members of the genus *Homo* were to some degree polygynous. . . . [T]he occurrence of multiple sexually dimorphic traits in *H. sapiens* (in body size, shape and composition, facial traits, secondary sexual adornments, the larynx and vocal pitch, and age at reproductive maturity) all indicate that human ancestors were polygynous.[12]

Dixson thus refers not only to gross sexual dimorphism—in height and body mass—but also to a variety of other, strongly suggestive traits such as the physical adaptations associated with a deeper male voice (characteristic of polygynous primates), as well as such facial traits as a beard and elongated chin, which exaggerate "maleness" and which, once again, are characteristic of nonhuman primates that are polygynous. These and other findings led biologist E. O. Wilson to conclude that human beings are "moderately polygynous,"[13] echoed by reproductive physiologist Roger Short: "We are basically a polygynous primate in which the polygyny usually takes the form of serial monogamy."[14]

Nonetheless, the situation, although very strongly suggestive, isn't yet altogether "proved."[15] Some specialists argue that insight can be gained by examining a seemingly irrelevant measure: the ratio of the length of the second digit to the fourth digit of the same hand. This appears to be an indicator of fetal androgenization—for reasons completely mysterious at present—with males having significantly lower ratios than females. This ratio tends to be lower in species that are more polygynous, perhaps because higher testosterone in males is selected for when male–male competition is intense. In any event, fossil digit ratios confirm that the prehuman evolutionary line was indeed polygynous, although with modern forms less so than earlier ones,[16] and the probably polygynous Australopithecines were outliers. A recent technical manuscript describing this situation was titled "Digit Ratios Predict Polygyny in Early Apes, *Ardipethecus*, Neanderthals and Early Modern Humans but not in *Australopithecus*."[17] Go figure.

> Finding 2: In all polygynous species, males aren't just larger than females, they are more prone to aggression and violence, especially directed toward other males. In many cases, males are also outfitted with intimidating anatomy that contributes to their potential success: horns, antlers, large canines, sharp claws, and so forth. But once again, these accoutrements only make sense insofar as their possessors are inclined to employ them.

It wouldn't do for a bull (elk, seal, or baboon), no matter how large and imposing, to refrain from using his bulk when it comes to competing with other bulls. There is little evolutionary payoff to being a Ferdinand among bulls. No matter how imposing, he could refrain from the competitive fray, save himself the time and energy his colleagues expend in threatening, challenging, and—if need be—fighting each other, not to mention the risk of being injured or killed in the process. Ferdinand would doubtless live a longer life and probably a more pleasant one. But when he dies, his genes would die with him. Publish or perish.

Accordingly, just as polygyny generates sexual dimorphism in physical size, it works similarly with regard to behavior and for the same basic reason. As with the male–female difference in physical size, male–female differences in violent behavior vary with the degree of polygyny. (Degree of polygyny can be operationalized more precisely by looking not simply at average harem size but at the ratio of male-to-female variance in reproductive success. In cases of true sexual monogamy, this ratio is 1. With increasing polygyny, the ratio mounts above 1; and with increasing polyandry, it dips below 1.) Among monogamous species, males and females are pretty much equal when it comes to violent inclinations because they are equal when it comes to the reproductive payoff of violence itself.

As expected, the male–female difference in aggressiveness and violence among highly polygynous species is very great. And what of moderately polygynous species? Here the difference is, not surprisingly, moderate.

Among human beings, men—starting as boys—are more aggressive and violence prone than are women. I have found that the male–female difference in perpetrators of violent crime is about 10 to 1, consistent across every state in the United States, and true of every country for which such data are available. Moreover, this difference is greater yet when proceeding from crimes that are less violent to more violent: the male–female difference in petty crime, for example, is very slight, greater when it comes to robbery, greater yet with regard to assault, and most dramatic in homicides. This is true even when the actual crime rates differ dramatically across different countries. Thus, the homicide rate in Iceland is about 1% that in Honduras, but the male:female ratio of those committing homicide is essentially unchanged. Overall cultural differences between Iceland and Honduras are very great, which doubtless explains the overall difference in homicide rates. However, male–female differences remain proportionately unchanged, just as male–female differences in human biology don't vary between Iceland and Honduras, or indeed any place people are found.[18]

Finding 3: In all polygynous species, females become sexually and socially mature at a younger age than do males. This fact is superficially counterintuitive because when it comes to reproducing, females by definition are the ones who bear the greater physiological and anatomical burden: eggs are much larger than sperm; among mammals, females, not males, are the ones who must build a placenta and then nourish their offspring from their own blood supply. Females, not males, undergo not only the demands of pregnancy and birth; they also provide all the calories available to their infants via nursing (and, little known to most people, lactation actually makes even greater energy demands than does gestation).

Based on these considerations, we would expect that if anything, females would delay their sexual maturation until they are proportionally larger and stronger than males, because when it comes to producing children, the biologically mandated demands on males are comparatively trivial: just a squirt of semen. But in fact not only are females typically smaller than males as we have seen, but they become sexually mature earlier rather than later because of yet another consequence of the power of polygyny. Male–male competition (mandated, as we have seen, for the harem-keeping sex) makes it highly disadvantageous for males to enter the competitive fray when they are too young, too small, and too inexperienced. A male who seeks to reproduce prematurely would literally be beaten up by his older, larger, and more savvy competitors; whereas early breeding females—who don't have to deal with the same kind of sociosexual competition—don't suffer a comparable penalty.

And so, among polygynous species, females become sexually and socially mature earlier than do males. The technical term is "sexual bimaturism," and anyone who has ever seen 8th, 9th, or 10th graders, or has simply been an adolescent, will immediately recognize the phenomenon whereby girls aged 12 to 16 are not only likely to be (temporarily) taller than their male classmates but also considerably more mature, socially as well as sexually.

Once again, and as expected, the degree of sexual bimaturism among animals varies directly and consistently with their degree of polygyny. Sexual maturation occurs at roughly the same age for males and females in primate species that are monogamous: for example, Latin American marmosets and owl monkeys. Among polygynous species such as rhesus macaques, squirrel monkeys and, indeed, nearly all primates (human as well as nonhuman), males mature more slowly. Thus they reach social and sexual maturity later than do females, when they are considerably older and larger than "their" females and also, not coincidentally, considerably older and larger than the other, less successful males. The sexual bimaturism so familiar to observers of Western teenagers (and to those teenagers themselves!) is a cross-cultural universal.[19]

Among the Aché of Paraguay (a hunting/foraging people who have been especially well studied by evolutionary anthropologists), "Teenage girls become women, and soon become the object of the attention of older men, whilst boys of the same age are still small, thin, and immature relative to the adult men."[20]

As if it were needed, here is further reason to have confidence in the significance of Findings 1, 2, and 3: among those exceedingly rare species that are polyandrous ("reversed polygyny" in which one female is socially and sexually paired with more than one male), the preceding correlations of sexual dimorphism in size, behavior, and age at maturity are predictably reversed, with females larger and more aggressive than males and becoming sexually mature later rather than earlier.

Among the very few polyandrous species are several types of birds, such as the spotted sandpipers of North America and some species of Latin American marsh dwellers known as jacanas or lily-trotters (because they have elongated toes that enable them to walk on top of water lilies). Males maintain small territories on the edge of a pond. Females are about twice as big, and they compete vigorously with other females, defending large territories that typically contain several males. Females normally lay four eggs in the nest of each of their "husbands," who then do all the incubation as well as caring for the precocious young after they hatch.[21] Polyandry is not found among any of the great apes and is exceedingly rare in mammals generally.

> Finding 4: Finally, in our perusal of the evidence for human polygyny, we have the simple historical record, confirmed by modern anthropology.[22] When it comes to mammals, monogamy is very rare, polyandry is almost unheard of, and polygyny is overwhelmingly the most common—although exact numbers are difficult to come by. And as for human beings, once again we are revealed to be perfectly good mammals.

The great French anthropologist and social theorist Claude Levi-Strauss long argued that marriage is a social and economic arrangement rather than a sexual or biological one. In a sense, he is correct, because as we have seen, there is every reason to believe that we are naturally polygynous, with monogamy imposed on our biology rather than emanating from it. Nonetheless, Levi-Strauss acknowledged that "even in a strictly monogamous society . . . [a] deep polygynous tendency . . . exists among all men."

"When it comes to polygyny," concurs anthropologist Weston La Barre,[23]

the cases are extraordinarily numerous. Indeed, polygyny is permitted (although in every case it may not be achieved) among all the Indian tribes of North and South America, with the exception of a few such as the Pueblo. Polygyny is common too in both Arab and Negro groups in Africa and is by no means unusual either in Asia or in Oceania. Sometimes, of course, it is culturally limited polygyny: Moslems may have only four wives under Koranic law, whereas the King of Ashanti in West Africa was strictly limited to 3,333 wives and had to be content with this number.

A cross-cultural survey of 849 societies found that prior to Western imperialism and colonial control over much of the world—which included the imposition of historically recent Judeo-Christian marital rules—708 (83%) of indigenous human societies were preferentially polygynous. Among these, roughly one-half were usually polygynous and one-half occasionally so. Of the remainder, 137 (16%) were officially monogamous[24] and fewer than 1% polyandrous.[25]

These numbers are potentially misleading. Notice that I wrote "preferentially polygynous" and not "exclusively polygynous." The fact that roughly more than three-quarters of human societies *preferred* polygyny does not mean that everyone— specifically, every man—within them was either a harem keeper or a bachelor. Rather, in nearly all cases polygyny coexists along with monogamy as well as bachelorhood. In fact, it appears that during most of human evolutionary history, most men ended up with one wife.

Detailed research on the Xavante people of Brazil provides an example. In a sample of 184 married Xavante men, 110 were monogamously mated (more precisely, they had one wife), whereas 74 had more than one. Some high-ranking men had four or even five wives; and not surprisingly, such men had two to three times more children than did other men of the same age. By age 40, 6% of Xavante men were childless; whereas by contrast, out of 195 women, just 1 was childless by age 20. The variance in reproductive success among Xavante men was 12.1; among women, 3.9.[26]

Genomic studies based on retrieved fossil human DNA has further confirmed our species' primordial polygyny.[27] Looking at the mid-range human fossil record, there is a disparity between the relatively low levels of Y-chromosome diversity (the Y chromosome is inherited from fathers only) and the greater diversity in mitochondrial DNA (which is inherited only from mothers). The researchers concluded that this is most parsimoniously explained by the likelihood that "until recently only a few men . . . contributed a large fraction of the Y-chromosome pool at every generation." In other words, our ancestors were mostly polygynous, with the effective breeding population of women being substantially greater than that of men—because whereas most women reproduced, a small number of men were disproportionately doing so. Then, over time, "the variance of their [the males'] reproductive success . . .

decreased, through a recent shift from polygyny to monogamy," as more men began contributing their own Y chromosomes to the gene pool. The researchers conclude— based purely on the DNA evidence—that "over much of human prehistory, polygyny was the rule rather than the exception."

Their estimates are that much of this transition from polygyny to increasing frequency of monogamy occurred between 10,000 and 5,000 years ago. Insofar as this is accurate, the fact that so many traditional societies remained polygynous even into modern times speaks strongly to our species-wide polygynous stubbornness. Our shift from polygyny to monogamy, in short, is a recent evolutionary development and is still a work in progress.

The word "monogamy" as used here—and pretty much everywhere—should be corrected to one wife *at a time,* since by far the most frequent pattern (for women as well as for men) has long been "serial monogamy," which could as well be stated as "serial polygamy." As I'll discuss later, human beings also show a powerful inclination for pair bonding (which nonetheless doesn't preclude a parallel penchant for polygamy). It is exceedingly rare for either a man or a woman to have literally one and only one sexual partner in his or her entire life. Pretty much all men strive to be polygynously mated, whether formally and legally, where permitted, or informally, simply as a result of their personal behavior (concubines, prostitutes, or infidelity)— resulting in de facto polygyny.

Polygyny, among men, and polyandry, among women, needn't have been universal to have had a profound evolutionary impact on our ancestors and thus on ourselves. Certain situations have an effect that transcend their statistical frequency. For example, actual occurrences of predation on human beings appear to have been relatively rare, as they clearly are today, yet the fear and "memory" of being victimized by large carnivores (or by comparatively small but lethal snakes) has left a large imprint.

Vestigial taints aren't limited to anatomy, as with the rudimentary skeletal "legs" of whales or perhaps the outmoded human appendix. Wildlife biologist John Byers has been impressed with the extraordinary running speed of North American pronghorns (often wrongly called "antelopes," they're actually closer to goats). Many of the pronghorn's traits—including aspects of its social organization and mating behavior—derive from its remarkable adaptation to high-velocity escape, which paradoxically exceed any current predation threat. The now-extinct American cheetah and its relatives used to streak over prairies now occupied only by more slow-moving coyotes, wolves, and cougars, and even slower grizzly bears.[28] Although, as Byers points out, "predator-driven selection has been relaxed," pronghorns remain speedy beyond their obvious need, haunted by "ghosts of predators past." We are similarly dogged by the ghosts of ancestral successes and failures, driven by polygamous pressures that are mostly (although not entirely) past as well.

When a biological impact is sufficiently great—as with a small number of men having a disproportionate number of mates—the consequence can be significant, not only statistically but also practically, even if the number of affected individuals remains a minority. Insofar as some men succeeded in acting on their polygynous propensities, and as a result generated an outsized return on their evolutionary balance sheet, this would have skewed the variance in reproductive success among men compared to that for women and continues to do so.

The fact that male reproductive competence continues into middle and even old age is also consistent with polygyny because the human species-wide pattern is that men accumulate wives as they age, whereas women do not accumulate husbands. (It is also noteworthy that in the Western world, where explicit polygyny is illegal, divorced middle-age men are more likely to remarry than are middle-age women.[29]) The bottom line is clear. When it comes to reproductive performance, men have long varied more than have women, and this greater variance is both a result of the underlying biology of maleness compared to femaleness (sperm making vs. egg making), as well as a cause of many other interesting consequences of being human.

Although evolutionary biologists have only in the past few decades come to appreciate the significance of polygamy and how it generates differences in the reproductive situations of men and women, recognition of its peculiar effects isn't nearly so new. To my knowledge, the first clear statement appeared in 1710, in an article published in the *Philosophical Transactions of the Royal Society of London*, by one Dr. John Arbuthnott, and titled "An Argument for Divine Providence, Taken from the Constant Regularity Observed in the Births of Both Sexes." In it, the author maintains that

> Polygamy is contrary to the Law of Nature and Justice, and to the Propagation of the Human Race: for where Males and Females are in equal number, if one man takes Twenty Wives, Nineteen Men must live in Celibacy, which is repugnant to the design of Nature; nor is it probable that Twenty Women will be so well impregnated by one Man as by Twenty.

Dr. Arbuthnott's arithmetic is correct, as is his assertion that polygamy is contrary to current Western conceptions of justice, although rather than being contrary to the "Law of Nature," it is—like it or not—fully consistent with that "law." As we shall see, there is reason to conclude that in most cases, those 20 hypothetical women really are ill-served by polygamy, but for reasons that are more complicated and more interesting than merely because they are likely to be less "well impregnated."

A t the risk of sounding Freudian, the truth is that just as the male sexual apparatus is more obvious and external than its female counterpart, men show readily apparent indications of their sexual evolution—not only in their anatomy but also their behavior—whereas women are more subtle about theirs. And yet, the biological reality is that human beings are both polygynous and polyandrous. This might sound oxymoronic, equivalent to a species being both big *and* small, aquatic *and* terrestrial, aggressive *and* nonaggressive, and so forth. But paradoxically, polygyny and polyandry can coexist. Male–male competition and male-based harem keeping (polygyny) is overt, readily apparent, and carries with it a degree of male–male sexual intolerance that also applied to polyandry, whereby "unfaithful" women along with their paramours are liable to be severely punished if discovered. This intolerance is easy enough to understand because the evolutionary success (the "fitness") of a harem-keeping male is greatly threatened by any extra-curricular sexual activity by "his" females; they could be inseminated by someone other than the harem keeper, which would result in a payoff for the paramour and a fitness decrement for the cuckolded male. As a result, selection has not only favored a male tendency to accumulate as many females as possible (polygyny) but also an especially high level of sexual jealousy on the part of males generally. And this, in turn, pressures polyandry into a degree of secrecy not characteristic of polygyny. Another way of looking at it: patriarchy pushes polyandry underground but does not eliminate it.

Female harem keeping—polyandry—goes against some aspects of human and mammalian biology, once again because of the difference between sperm making (what males do) and egg making (a female monopoly). Although a male's fitness is enhanced with every female added to his mating prospects, the same is much less true for the fitness of a female who mates with additional males. As we shall see, there can indeed be a payoff to females who refrain from sexual exclusivity (actu-ally, there are many such payoffs); however, there are also substantial costs, not the least of which is running afoul of the male sexual jealousy just described. As a result, even though females can sometimes enhance their fitness by mating with additional males, they are simultaneously selected to be surreptitious about their sexual adven-turing. Hence, polyandry—unlike its overt counterpart, polygyny—is more likely to be covert and hardly ever proclaimed or institutionalized. It also doesn't reveal itself in such blindingly obvious traits as sexual dimorphism in physical size, aggressive-ness, or differences in age at sexual maturity because unlike the situation among males, natural selection does not clearly reward such readily apparent traits among females.

Men, in their befuddlement, have had a hard time seeing female sexuality for what it is, consistently either over- or underestimating it. Thus, women have often been portrayed as either rapacious and insatiable, or as lacking sexual desire altogether. At one time, Talmudic scholars entertained such an overblown estimate of women's sexuality (and society's responsibility to repress it) that widows were forbidden to keep male dogs as pets! But as noted psychologist Frank Beach pointed out, "any male who entertains this illusion [that women are sexually insatiable] must be a very old man with a short memory or a very young one due for a bitter disappointment." Or, as anthropologist Donald Symons put it, "The sexually insatiable woman is to be found primarily, if not exclusively, in the ideology of feminism, the hopes of boys, and the fears of men."

At the other extreme, an influential 19th-century Victorian tract by Dr. William Acton announced that "the majority of women (happily for society) are not very much troubled with sexual feelings of any kind. What men are habitually, women are only exceptionally."[30]

Even today, a woman's sexuality is often measured by her ability to arouse desire in someone else rather than to experience it herself. These distortions—by men—arise largely because they view female sexuality as threatening. The threat arises not only because a fully sexual woman evokes deep fears of inadequacy (especially inability to satisfy such a woman) but also because female sexual independence and initiative threaten the tenuous thread by which a man can have confidence that his wife's children are his own. When not irrationally jealous and/or nervous, therefore, men are nonetheless often comforted by the myth of the unsexed woman, even if deep in their hearts they know it is a lie.

In this regard, it is worth mentioning that some anthropologists have recently begun reassessing the received wisdom as to polyandry's rarity. A new category, "informal polyandry," has been proposed to include societies in which more than one man mates regularly with the same woman.[31] These circumstances are found in a number of societies beyond the standard "classical polyandry" of the Himalayas, Marquesa Islands, and parts of the Amazon basin. Informal polyandry often co-occurs with a local belief system known as "partible paternity" in which it is thought that if multiple men (albeit rarely more than two or three) have sexual intercourse with a pregnant woman, then they literally share paternity of any offspring that result. More about partible paternity in chapter 4, but for now let's note that a number of hypotheses have been offered to explain the existence of informal polyandry, rare as it is—because in theory, it should be more rare yet.

Of these, the one most consistent with available data is that informal polyandry is especially likely when the sex ratio is unbalanced so that there are substantially more men than women. This could be caused by preferential female infanticide, which generates an excess of men, or if polygyny is simultaneously practiced in the same society. With a small number of men monopolizing more than their arithmetic share of women, a number of men are left out of the marriage market, which may in turn lead them to elect to "share" a wife rather than be sexually and reproductively excluded altogether. Better a fraction of a wife than none at all.

Among the Inuit, for example, "exceptionally great hunters are able to support more than one wife; good hunters can support one wife; and mediocre hunters, or those unwilling or unable to take a wife from another man, share a wife. In such instances co-husbands are making the best of a bad situation and perhaps staying alert for new marital opportunities."[32] The preceding focuses on the male perspective, appropriately so because male reproductive success is unavoidably threatened by polyandry; at least, it is lower than for successful polygynists or even monogamists. However, informal polyandry also deserves to be seen as a strategy for and by women, activated when the sex ratio is such that they are in short supply and can drive a hard bargain, essentially forcing men into a reproductive arrangement they would not otherwise choose. In any event, it wouldn't occur if women weren't, at least on occasion and in certain circumstances, polyandrous.

It would be interesting to see if something comparable happens in modern Western environments in which men greatly outnumber women: for example, in colleges and universities that were traditionally woman only and are just beginning to admit men, or in military installations in which men greatly outnumber women. In any event, in this book, I focus somewhat more on polygyny and its evidence and consequences for men than on polyandry and its evidence and consequences for women, simply because at this point in our scientific knowledge, we know more about the former. Granted that the evidence for human polyandry is more speculative than dispositive, here, nonetheless, is a sample of the arguments:

Argument 1: Human beings are unusual among mammals in that females conceal their ovulation. In most species, ovulation is conspicuously advertised, but not in us! Indeed, even in the medically sophisticated 21st century, and despite the fact that reproduction is such a key biological process, it is remarkably difficult to ascertain when most women are ovulating. There is considerable controversy over why women's ovulation is kept so secret, but one intriguing possibility is that it facilitates a woman's ability to deceive her mate as to precisely when she is fertile. If our great-great-grandmothers sported a dramatic pink cauliflower bottom when they were in season, our great-great-grandfathers could have mated with them at such times and then ignored them while pursuing their own interests. (Which notably would have included mating with other women.) But by hiding their ovulation, our female ancestors essentially made it impossible for our male ancestors to guard them all the time, giving the option for great-great-grandma to sneak copulations with other men, of her choosing, thereby avoiding the ire of her social partner while obtaining whatever benefits such "extra-pair copulations" may have provided. More on this later.

Argument 2: Women are also unusual among mammals in lacking a clear behavioral estrus (or "heat") cycle. As a result, they are able to exert a remarkable degree of control over their choice of a mate, unlike most female mammals, who find themselves helplessly in thrall to their hormones. Absent such choice, polyandry would be indistinguishable from literal promiscuity. The word "promiscuity" carries with it a value judgment, distinctly negative. For biologists, however, it simply means the absence of subsequent bonding between the individuals involved. Some animals appear to be truly promiscuous. For example, many marine organisms—such as barnacles or oysters—squirt their gametes (eggs and sperm) into the water, usually responding to chemical signals but not engaging in anything like mate choice. But for the most part, promiscuity is rare because nearly all living things—females in particular—are more than a little fussy when it comes to settling on a sexual partner, even as they may end up with many such partners.

Females in general and women in particular have a substantial interest in making a good mating choice (or choices), if only because biology mandates that such decisions are especially consequential for them given that children are born quite helpless, and their prospect of biological success in enhanced by many factors, notably parental care and attention, in addition to good genes. And indeed, females in general and women in particular are especially fussy when it comes to choosing their sexual partner(s).

Argument 3: Recent studies by evolutionary psychologists have shown that during their ovulatory phase, women are attracted to men whose body build and facial features reflect high testosterone and basic good health (i.e., good genes); whereas otherwise, they are more influenced by indications of intelligence, kindness, sense of humor, ambition, and personal responsibility. In other words, women follow a two-part reproductive strategy consistent with an evolutionary history of polyandry: mate, when possible, with partners carrying those good genes, but associate socially with those offering the prospect of being good collaborators and co-caretakers of children. For some women—those fortunate to pair with men providing both good genes *and* good parenting/protection prospects—the risks of polyandry (especially, their husband's ire, potential violence, and possible abandonment) outweigh the possible benefits. But for others—quite possibly the majority—the opposite can be true.

Argument 4: The adaptive significance (evolutionary payoff) of female orgasm has long been debated. Among the possible explanations—all consistent with polyandry—is that orgasm enables women to assess the appropriateness of a short-term mating partner as a long-term prospect, while another suggests that female climax is not only rewarding for the woman in question but also reassuring for her partner, providing confidence that she will be sexually faithful, while giving her the opportunity to be exactly the opposite.

Argument 5: Given that polyandry among animals is predictably correlated with reversals in the "traditional" forms of sexual dimorphism, why hasn't human polyandry resulted in women being larger and more aggressive than men? For one thing, it isn't possible for men to be larger than women (which, as we've seen is mostly a result of polygyny) *and* for women to be larger than men (because of polyandry)! And for another, because of the difference between sperm and eggs, the reproductive payoff to polygyny—and its associated male–male competition—is substantially greater than that of polyandry, which in turn has caused polygyny to be the more prominent driver of human sexual evolution. This is not to claim that polygyny (acting mostly on males) is any more real than is polyandry (acting mostly on women), but rather that its effects are more dramatic and more readily identified.

Argument 6: Because of the negative fitness consequences for men resulting from polyandry on the part of "their" women, we can expect that men would have been selected to be quite intolerant of it. And indeed, sexual jealousy is a pronounced human trait—also widespread among animals—not uncommonly leading to physical violence. Such a powerful and potentially risky emotional response would not have been generated by evolution if women weren't predisposed, at least on occasion, to behave in such a way that is adaptive for them.

Female sexual jealousy can also be predicted, and it certainly occurs, but for biologically comprehensible reasons it is notably less intense, especially because male sexual dalliance doesn't threaten the fitness of the "scorned wife" nearly as much as comparable behavior by wives threatens the fitness of cuckolded males.

"Mommy's babies," goes the saying, "Daddy's maybes." Women don't need to worry whether their children are genetically theirs. Men have no comparable confidence, so evolution has outfitted them with more sexual intolerance. Women can nonetheless be threatened by male gallivanting, especially since this might result in loss of their mates' personal devotion and at least some of his investment in their children. But this is quite different from the male situation, whereby a single out-of-pair copulation by a female can result in a catastrophic fitness decrement for the deceived male.

Not surprisingly, therefore, the patriarchal "double standard" is pretty much a cross-cultural universal. Once again, this is because female infidelity raises the possibility of a man rearing another's children, whereas its male counterpart—for all the problems it may cause to the relationship—doesn't necessitate a comparable cost of cuckoldry.

To get a clear understanding of a general process, it often helps to pay special attention to the extremes. This is particularly true of a spectrum phenomenon, of which polygyny is an excellent example (as explained previously, there are varying "degrees of polygyny" just as there are degrees of polyandry). Accordingly, let's take a quick look at an extreme case of polygyny in the animal world because when we do, we'll see ourselves—albeit in caricature.

Elephant seals are very, very large: in fact, elephantine. Bulls can reach 16 feet in length and weigh more than 6,000 pounds. Cows are much smaller, about 10 feet long and weighing around 2,000 pounds. This size difference is important, arising as it does because of the elephant seal mating system: the species might be the most polygynous of all mammals, with successful males establishing harems of up to 40 females. Because (as in most species) there are equal numbers of males and females, this means that for every highly successful bull seal, there are roughly 39 unsuccessful, reproductively excluded bachelors. In the world of elephant seals, every healthy female gets mated, but only a very small proportion of males are comparable evolutionary winners. On average, 4% of the bulls sire 85% of all the offspring.[33] Bulls therefore fight long and hard among themselves for possession of a harem. Success requires large size, a violent temperament, massive canine teeth combined with willingness to employ them, a thick chest shield to provide protection from one's opponent, and sufficient age and experience.

Female elephant seals wean their babies in late summer and early fall, after spending much of the summer on land, members of a crowded, beachfront harem. It turns out that by the time they are weaned, some young elephant seals are considerably larger than others—as much as twice the size of their fellow weanlings. These oversized juveniles are known as "super-weaners." Their greater size conveys a distinct benefit because after spending a more or less idyllic time on their rocky beaches, nursing from their mothers, at summer's end and upon being weaned, the pups must begin a long sojourn at sea, not returning to land until the following spring. This is, not surprisingly, a stressful time for young elephant seals, and—also not surprisingly— those who were super-weaners are more likely to survive. It isn't known whether male super-weaners are, in turn, more prone to eventually become harem masters, but it's a good bet because in a highly competitive system, anything likely to provide a "leg up" when it comes to physical condition is likely to bring benefits.

So far, so good, at least for the super-weaners. A question arises, however. Why—given the payoff of being supersized—aren't all elephant seals super-weaners? It turns out that because elephant seal mothers are limited in how much milk they can produce, there is only one way to become a super-weaner: a pup must obtain milk from two lactating females. How is this achieved? It's not easy. Females are quite determined to make their milk available only to their offspring, not to someone else's. This selfishness makes a lot of evolutionary sense because nursing mothers who were profligate with their precious milk would have left fewer descendants (and thus, fewer copies of their milk-sharing genes) than others who were disinclined to wet-nurse an unrelated pup.

Nonetheless, even though every pup has only one genetic mother—"partible maternity" being at least as unrealistic as "partible paternity"—it's still possible for a pup to get milk from two "mothers." Elephant seal pups occasionally die while nursing, either from "natural causes" or because they are literally squashed during the titanic battles among oblivious, competing bulls, who have females on their mind, and not the safety of young pups, who were sired the previous year, possibly by a different male. The death of nursing infants provides an opportunity for an enterprising young pup: if he can locate a bereaved mother—quickly enough after her infant has died so that her milk hasn't dried up—he might induce her to permit him to nurse in place of the recently deceased infant.

This is an effective strategy, but also a risky one, because most females don't take kindly to allowing an unrelated baby to suckle. "Sneak sucklers" often get bitten and may die of their wounds. But successful ones become what are known (in the technical literature, thanks to the detailed research of elephant seal maven Burney Le Boeuf) as "double mother suckers," and they, in turn, become super-weaners. Here is the kicker: *all* double mother suckers are male! Chalk it up to the pressure of polygyny, in the case of elephant seals, super-polygyny leading—because of the potential payoff to males of being larger, stronger, and healthier than their competitors—to super-weaners by way of double mother sucking. All of this requires, of course, a willingness to take risks, certainly greater willingness than is shown by female pups, who, as the harem-kept sex rather than the harem keepers, are pretty much guaranteed the opportunity to breed so long as they survive. For males in a highly polygynous species, mere survival isn't enough. They must stand out from their peers.

As we'll see, a number of human traits can be understood as resulting from our shared human history of moderate polygyny. Human beings aren't elephant seals. Few—if any—of our fellow *Homo sapiens* are double mother suckers. Nonetheless, the data are overwhelming that little boys are more risk-taking, on average, than are little girls,[34] a difference that continues throughout life and is most intense among adolescents and young adults—precisely the age at which reproductive competition was most intense among our ancestors, and to some extent, still is today. Examples of extreme polygyny, such as elephant seals, reveal exaggerations and caricatures of traits found in human beings as well. We are biologically designed to be mildly, not wildly, polygynous; but those traits found in such extreme cases as elephant seals, elk, and gorillas shed light on the more modest but nonetheless real and otherwise perplexing reality of what it means to be human.

In attempting to unveil that reality, I will often refer to what anthropologists have discovered about the world's traditional societies. These findings are especially interesting for any effort to penetrate to the core of human nature, but not because any such group is in any sense ancestral to the rest of us. All are equally human, which, for biologists, simply means that we are all capable of exchanging genes and producing viable offspring. On the other hand, it is clear that our ancestors spent upward of 99% of their—our—evolutionary past living in relatively small social groups and with comparatively little advanced technology, at least by modern Western standards. As Jared Diamond has emphasized, these people can be especially revelatory when we realize that they represent, in some sense, the "world until yesterday,"[35] a world that still whispers within all of us and that we can glimpse if we recognize that whereas no currently extant society represents our ancestors, there is much that we can discern if we look and listen carefully. Patterns that speak to what anthropologists call "cross-cultural universals" are especially portentous.

Here is another way of looking at it. The diversity of human societies provides what is essentially a vast, worldwide experiment in which one thing—basic human biology—is held constant (after all, we are all members of the same species), while cultural traditions vary greatly—literally, all over the map. If, under these conditions, certain patterns emerge persistently, it is reasonable to consider that they may well be due to the one thing they all share: our human nature, as members of the species *Homo sapiens*. This makes the testimony from diverse human societies especially relevant, whether these societies are traditional or "modern."

A further benefit of taking traditional societies seriously is that it widens our perspective beyond the rather limited sample often studied by social psychologists, sociologists, and economists. For reasons of convenience, most laboratory and survey research has involved people from Western industrialized societies; worse yet, the most common research subjects are college sophomores, an even more limited subset. The disparity between the kinds of people who have been studied and the diversity of those who actually inhabit our planet, and who represent the range of inclinations found in the species *Homo sapiens*, has led researchers to designate—and to some extent, denigrate—such research as being overwhelmingly oriented toward people who are White, Educated, and from Industrialized, Rich, and Democratic countries: in short, they're WEIRD.[36]

And so, dear reader, if you find yourself wondering why this book devotes so much attention to seemingly exotic people such as the Hadza, Inuit, Xavante, Yanomamo, and so forth, please bear in mind that *you're* the ones who are WEIRD! Or, more accurately, if we want a realistic understanding of the diversity as well as the preferences of human beings more broadly, we had better consider a broad swath of these human beings—in addition to the fact that a case can be made that non-technological, traditional societies more closely approximate at least some of the conditions in which most of our ancestors spent most of their time.

That said, most of the currently available, large-sample, in-depth research involves a limited sampling of WEIRD people—if only because they are the ones readily available to researchers, most of whom, let's face it, are WEIRD themselves. Insofar as the available results are what we currently have to work from—particularly with regard to various male–female differences in sexual preferences—they, too will be reported in the following pages. It can be hoped that such studies will become more "cross-cultural" over time, although the reality is that most traditional societies are currently transitioning to more or less "modern" lifestyles. It is unlikely that these lifestyles reflect our natural human condition. This, by itself, adds further urgency to the effort to understand what human beings "really are," before our species is culturally steamrollered into homogeneity.

A s already noted, this book will doubtless be controversial, and not only among those who maintain that "human nature" (along with much of what's left of objective reality) is merely a social construct. It will also arouse the ire of those who argue that all male–female differences in behavior or social situation are attributable to culture, with biology being fundamentally irrelevant. Like it or not, however, biology matters. This is especially true because of the two primary biological factors that distinguish males from females, and thus, men from women: the fundamental difference between being a sperm maker and an egg maker, and, in the case of species that engage in internal fertilization, the basic disparity between males—who lack confidence when it comes to genetic connectedness with "their" offspring—and females, for whom such connectedness is guaranteed. And this remains true despite the fact that for some individuals, the boundaries between male and female are remarkably indistinct.

For example, the 2003 Pulitzer Prize in fiction went to the novel *Middlesex*, which describes the experiences of someone with an unusual, genetically based biochemical anomaly: 5-alpha-reductase deficiency, or late onset virilization. The main character began life identified as a female; then, at puberty, he developed distinctly male traits. There are many such cases, sometimes involving an individual who is seemingly "normal" when it comes to genes and biochemistry, but who nonetheless is determined that his or her body does not reflect the underlying "real" self, to the degree that hormone treatments and sex reassignment surgery are necessitated. The seemingly simple distinction between male and female is yet more confusing when it comes to the conundrum of XY females and XX males, which reverses the usual pattern. Here, the genetic phenomenon of crossing over—whereby chromosomes exchange material prior to meiosis when eggs and sperm are produced—results in an important gene being either added or deleted. This gene activates some gender-influencing genes and inhibits others so that someone can end up with a Y chromosome, yet display many otherwise female traits, or can have two Xs, and yet appear (mostly) male.

It is especially admirable that in recent years, not just physicians but the public has become increasingly aware that the simple dichotomy of male or female doesn't begin to exhaust the existence of numerous intermediate conditions. For example, individuals with Turner's syndrome have one X chromosome but no Y; that is, their sex chromosomes can be designated XO. Because the sex-determining DNA appears on the Y chromosome, individuals lacking a Y chromosome develop into females; but because they have only a single X, Turner's individuals are deficient in estrogen and need supplementary hormone treatment. Turner's appears in roughly 1 in 3,000 live births. Y0 individuals don't survive.

Then there is Klinefelter's syndrome, which occurs when someone is born XXY. It occurs in approximately 1 in 600 live births. Klinefelter's individuals appear male because of the masculinizing effect of their Y chromosome. They are comparatively tall, but may also have some breast development. XYY individuals—who represent roughly 1 in 850 live births, are also tall, and may have impaired cognitive function, somehow related to the imbalance of being male with an extra Y chromosome. The list goes on. Congenital adrenal hyperplasia (CAH) is exceptionally rare—1 in 16,000 live births—and involves exposure to high levels of androgens since the adrenal glands make androgens along with other hormones. When CAH occurs in males, the effect is minimal; in females, it is behaviorally masculinizing, including more rough-and-tumble play. Adrenal insensitivity syndrome is also rare, about 1 in 10,000 live births. It is notable in XY individuals (who are therefore genetically male) whose androgen receptors are nonfunctional or dysfunctional. People with adrenal insensitivity syndrome look like females, develop breasts at puberty, have female genitals, seem female, but are genetically "male." They lack internal female reproductive organs: no uterus, Fallopian tubes, or ovaries.

In addition, of course, there are the better known examples of "gender bending" stereotypes: males who—because of their biological makeup, in ways that are not currently understood—choose to have sex with males, and females with females. Reproducing, however, is another story. Same-sex preference doesn't negate being male or female, which is biologically defined as being a sperm maker or egg maker, no matter one's erotic or romantic inclinations. The same applies to presence or absence of beards, breasts, or genitals: although gender—one's social role, as identified male or female—is fascinating and certainly noteworthy, it doesn't force reconsideration of the biological meaning or evolutionary consequences of sex, that is, being male or female. The fact that, in addition, some individuals aren't easily categorized as either male or female, doesn't make maleness or femaleness any less useful or important, or biologically meaningful.

There are times of day that cannot readily be categorized as either "day" or "night," such that our appreciation of nature's nuance is enhanced by paying attention to various intermediate situations that we label dawn and dusk. But the existence of these in-between situations doesn't reduce the meaningfulness of day and night: which, most of the time, are as different as, well, day and night. We use words to enable us to speak (and think) accurately about the world around us, and even though there is always a risk that we will mistake the signifier—for example, "man" or "woman"—for the thing signified, it remains true that there are genuine "things" out there in the real world, things that warrant being identified, even if the reality is occasionally more slippery than many people might wish.

On the other hand, we are most free from evolutionary constraints in proportion to our understanding of the inclinations and predispositions with which our biology has endowed us, or stuck us. I would like to think that the book you are reading treads firmly on factual ground, and is not a work of male chauvinism or of blinkered biology. In any event, I hope to clearly demarcate fact from speculation and to identify substantial realms of human freedom and individual possibility, offering the potential of transcending much of our biology: but only if we dare to know.

Notes

1. With the exception of sharing some intriguing hypotheses here and there, which are clearly labeled as such.
2. Barath, D. P., & Lipton, J. E. (2001). *The myth of monogamy: Fidelity and infidelity in animals and people.* New York: W. H. Freeman. [Also available in paperback: New York: Henry Holt, 2002.]
3. To some extent, the writing of any book is an exercise in arrogance.
4. Hrdy, S. (1986). Empathy, polyandry, and the myth of the coy female. In R. Bleier (Ed.), *Feminist approaches to science.* New York: Pergamon Press.
5. Variance is a technical term, defined in statistics. For our purposes, however, it is sufficient to think of variance as equivalent to variability, the extent to which a group of numbers is spread out. If every female has the same number of offspring, the variance in female breeding is zero; whereas if one male has 100 offspring and 99 have none, the variance in male breeding is very large indeed.
6. Of course, in a sense, this isn't really true because the overall payoff for both males and females is determined by the overall consequence of all fitness considerations, those involving ecological efficiency plus sexual competition. It's just that because sexual competition looms large when it comes to male fitness, their body size is shifted to the larger end; whereas females—insofar as they are insulated from substantial female–female competition—are more strongly selected to match the ecological optimum for their species.
7. Hrdy, S. B. (1981). *The woman that never evolved.* Cambridge, MA: Harvard University Press.
8. Plavcan, J. M. (2012). Sexual size dimorphism, canine dimorphism, and male-male competition in primates: Where do humans fit in? *Human Nature, 23,* 45–67.
9. Lovejoy, C. O. (2009). Reexamining human origins in light of *Ardipethecus. Science, 326,* 74–77.
10. Clarys, J. P., Martin, A. D., & Drinkwater, D. T. (1984). Gross tissue weights in the human body by cadaver dissection. *Human Biology, 56,* 459–473.
11. Puts, D. (2010). Beauty and the beast: Mechanisms of sexual selection in humans. *Evolution and Human Behavior, 31,* 157–175.
12. Dixson, A. F. (2009). *Sexual selection and the origins of human mating systems.* Oxford, England: Oxford University Press.
13. Wilson, E. O. (1978). *On human nature.* Cambridge, MA: Harvard University Press.
14. Short, R. V. (1994). Why sex? In R. V. Short and E. Balaban (Eds.), *The differences between the sexes.* Cambridge, MA: Cambridge University Press.

15. Philosopher of science Karl Popper maintained that scientific propositions can never be proved, but only disproved, with the capacity for disproof being one of the defining characteristics of a legitimate scientific hypothesis. On a Popperian basis, we must conclude that mild but genuine human polygyny has thus far resisted any disproof.

16. Nelson, E. C., Rolian, L., Cashmore, L., & Shultz, S. (2011). Digit ratios predict polygyny in early apes, *Ardipethecus*, Neanderthals and early modern humans but not in *Australopithecus*. *Proceedings of the Royal Society of London, B 278*, 1556–1563.

17. Ibid.

18. An evolutionary approach to human behavior is sometimes accused of being potentially or even essentially racist because of its concern with the role of genes. The truth, I maintain, is precisely the opposite: namely, that it is a potent *antidote* to racism because "human sociobiology" looks for cross-cultural universals, characteristics that all people share (regardless of their social traditions, skin color, etc.) by virtue of being members of the same species, and thus, brothers and sisters under the skin.

19. For example, Hill, K., & Hurtado, A. M. (1996). *Ache life history.* New York: Aldine.

20. Ibid.

21. Jenni, D., & Collier, G. (1972). Polyandry in the American Jacana (*Jacana spinosa*). *The Auk, 89*, 743–765.

22. Here I must note a problem. Anthropologists know far more about marriage rules than about actual mating behavior, so when a society is identified as "monogamous" or "polygynous," considerable uncertainty—or even outright skepticism—should be evoked. That said, it is far more likely that societies called monogamous actually practice covert polygyny than that polygynous people are actually secret monogamists. So if anything, estimates of polygyny are likely to err on the low side and vice versa for monogamy.

23. La Barre, W. (1954). *The human animal.* Chicago: University of Chicago Press.

24. As it happens, these proportions are almost identical to that found among primates, too: monogamy occurs in roughly 15% of nonhuman primate species, and various degrees of polygyny are found in nearly 85%, with polyandry virtually unknown.

25. Murdock, G. P. (1967). *The ethnographic atlas.* Pittsburg, PA: University of Pittsburg Press.

26. Salzano, F. M., Neel, J. V., & Maybury-Lewis, D. (1967). I. Demographic data on two additional villages: Genetic structure of the tribe. *American Journal of Human Genetics, 19*, 463–489.

27. Dupanloup, I., Pereira, L., Bertorelle, G., Calafell, F., Prata, M. J., Amorim, A., & Barbujani, G. (2003). A recent shift from polygyny to monogamy in humans is suggested by the analysis of worldwide Y-chromosome diversity. *Journal of Molecular Evolution, 57*, 85–97.

28. Byers, J. (1998). *American pronghorn: Social adaptations and the ghosts of predators past.* Chicago: University of Chicago Press.

29. Chamie, J., & Nsuly, S. (1981). Sex differences in remarriage and spouse selection. *Demography, 18*, 335–348.

30. Acton, W. (1862). *The functions and disorders of the reproductive organs in childhood, youth, adult age, and advanced life: Considered in their physiological, social, and moral relations* (3rd ed.) London: Churchill Publishers.

31. Starkweather, K. E., & Hames, R. (2012). A survey of non-classical polyandry. *Human Nature, 23*, 149–172.

32. Ibid.

33. Le Boeuf, B. J., & Reiter, J. (1988). Lifetime reproductive success in northern elephant seals. In T. G. Clutton-Brock (Ed.), *Reproductive success*. Chicago: University of Chicago Press.

34. Maccoby, E. E., & Jacklin, C. N. (1974). *The psychology of sex differences*. Stanford, CA: Stanford University Press.

35. Diamond, J. (2013). *The world until yesterday: What can we learn from traditional societies?* New York: Penguin.

36. Henrich, J., Heine, S. J., & Norenzayan, A. (2010). Beyond WEIRD: Towards a broad-based behavioral science. *Behavioral and Brain Sciences, 33*(2–3), 111–135.

2

Violence

Here is a startling truth. If we were ever to get rid of male-generated violence, we would pretty much get rid of violence altogether. What we might call the "killing establishment"—soldiers, executioners, hunters, even slaughterhouse operators—is overwhelmingly male. Underworld killers, such as violent gangs, are also peopled largely by men. From rampages in schools, movie theaters, and post offices to terrorist killings around the world, men—not women—are overwhelmingly the mass murderers. When people kill and maim other people, *men* are nearly always the culprits—and the victims too (although to a somewhat lesser extent).

The same applies to the uncountable private episodes of violence that receive little national attention but are the stuff of many a personal tragedy. Admittedly, an occasional Lizzie Borden and her ilk surfaces, but for every Bonnie there are about a hundred Clydes. Male brutalizers and killers are so common they barely make the local news, whereas their female counterparts achieve a kind of fame. A man who kills—even his own children—gets comparatively little notice, whereas when a mother kills her husband or offspring, it is news. For a man to generate a comparable response, his crime must be especially dreadful—such as serial murderer Ted Bundy, mass murderer Anders Breivik, cannibal Jeffrey Dahmer—or he must be a celebrity such as accused murderer O. J. Simpson. Violence may or may not be as American as apple pie, but it is as male as anything can be.

If we were ever to interrogate a hypothetical intelligent fish and ask it to describe its surroundings, one thing we almost certainly would *not* learn is that "It's awfully wet down here." Like our presumed piscine friend, people don't notice circumstances that are so widespread that we take them for granted.

The overwhelming maleness of violence is so pervasive in *every* human society that it is typically not even recognized as such; it is the ocean in which we swim. Another way of saying this: the fact that men are overwhelmingly the perpetrators of violence seems so natural that it takes real effort to appreciate how notable, how downright *strange*, it is. William James (philosopher as well as the first American professor of psychology) marveled at how difficult—yet important—it is to achieve sufficient distance from our own inclinations to make the "natural seem strange." Thus, he asked

> Why do we smile, when pleased, and not scowl? Why are we unable to talk to a crowd as we talk to a single friend? Why does a particular maiden turn our wits

so upside-down? The common man can only say, Of course we smile, of course our heart palpitates at the sight of the crowd, of course we love the maiden, that beautiful soul clad in that perfect form, so palpably and flagrantly made for all eternity to be loved! And so, probably, does each animal feel about the particular things it tends to do in the presence of particular objects. . . . To the lion it is the lioness which is made to be loved; to the bear, the she-bear. To the broody hen the notion would probably seem monstrous that there should be a creature in the world to whom a nest full of eggs was not the utterly fascinating and precious and never-to-be-too-much-sat-upon object which it is to her. Thus we may be sure that, however mysterious some animals' instincts may appear to us, our instincts will appear no less mysterious to them.[1]

It would be lovely if human violence were truly strange—that is, if it were so rare as to be outside normal experience and expectation. But in fact, it's pervasive. Also pervasive—and only slightly less taken for granted—is the reality that men are responsible for nearly all of it. To understand this phenomenon, we need to follow James's advice and make the naturalness of male-perpetrated violence seem strange enough that we can focus on it.

Few phrases have become as hackneyed as "battle of the sexes." And yet, male–female conflicts are insignificant compared to the genuine, violent battles that often take place *within* the sexes, almost all of them between males. Another overused yet underappreciated phrase is "angry young men," first applied to John Osborne's 1956 play, *Look Back in Anger*, and then associated with a generation of writers disillusioned with their place in British society. But in fact, young men needn't be British to be angry and therefore violent, and polygyny is much of what they are angry about.

Violence is a national (and worldwide) problem, and by and large, it is something that men direct toward other men. As with inner-city crime, in which both the perpetrators and victims are disproportionately minorities, men are disproportionately both the perpetrators and the victims of their own violence. I do not mean to romanticize or idealize women, or to deny that they too can sometimes be nasty, brutal, even violent and deadly. But when it comes to violence, the two sexes simply are not in the same league.

Hans Morgenthau, one of the great figures in 20th century political science, used to argue that politics was based on male competition for power, a competition that was, in turn, driven by three urges: to live, to propagate, and to dominate. Correct as far as he went, Professor Morgenthau might have been interested to learn that the first and third urges he identified are themselves proximate means to the middle one, the one that counts biologically: propagation. Reproduction, after all, lies at the root of *why* living things live and *why* they seek to dominate. The ultimate power of propagation explains why males in particular are often so eager to dominate, occasionally carrying their eagerness to violent extremes. We should not be surprised to find that aggressiveness is widely—and all too correctly—seen as manly and its alternative, timidity, as womanly. (When he was told that a high-ranking member of his administration had become a dove on Vietnam, Lyndon Johnson snarled "Hell, he has to squat to take a piss.")

Human violence is as multifactorial as anything in our repertoire, making it fool-hardy to suggest that there is just one cause. Not all violence derives from our polygynous promptings, nor is it all initiated by men; nonetheless, men are responsible for considerably more than their "fair share," just as they are disproportionately the victims as well. All of this makes polygyny a prime culprit. Like the energy captured in a bow that is drawn back, ready to power an arrow, polygyny generates much of the background tension that erupts into violence.

Not only are men on average 20% larger than women, a difference that we can with confidence attribute to polygyny-generated male–male competition; but women are to some extent directly responsible for this sexual dimorphism: around the world, when given the choice, women prefer taller men to shorter ones.[2] Both men and women, on average, would like their partners be taller than the average—bringing to mind the Lake Woebegone phenomenon in which "all the children are above average." A study titled "Separating Fact From Fiction: An Examination of Deceptive Self-presentation in Online Dating Profiles," found that men are more likely than women to lie about their height, testimony to men's recognition of women's preference.[3] In addition to being discouraging to people—men especially—who are vertically challenged, this finding adds another wrinkle of causation to the pattern whereby polygyny correlates with sexual dimorphism: enhanced male size may result not only from greater success in male–male competition but also from women's preference for larger men. This preference in turn could be due to the increased ability of larger men to defend females and their offspring, as well as to the "sexy son effect," about which there is more in chapter 3.

There are several different payoff prospects for polygynous—and would-be polygynous—males, all inclining them in the same direction: (1) Achieve access to multiple females, (2) Retain this access once achieved, (3) Avoid being a Darwinian loser, and (4) Be capable of defending your mate(s) and offspring. There might well be a fifth payoff, insofar as women (even if just some women) actively prefer males with those traits likely associated with successful polygyny. Of course, there are costs, too, notably the downside of being beaten up or even possibly killed, as well as "aggressive neglect," the intriguing phenomenon in which hyperaggressive males, because they are preoccupied with their various battles, risk paying insufficient attention to their mates and offspring.

Hence, even in the most polygynous situations, males have been selected to be strategists rather than gung-ho, risk-it-all daredevils. Consider sexual bimaturism, described earlier, in which males of polygynous species don't enter the competitive fray until they are large, strong, and savvy enough to stand what is literally a fighting chance.

Amid all this aggression and violence—and there is more to come—it's reassuring to note that in a now-famous study of mate preferences in 37 different cultures worldwide, the highest ranked out of 13 traits was "kind and understanding."[4] And this was true for both men and women.

This shouldn't, however, obscure the regrettable role of male aggressiveness and violence, which is apparent from even a superficial look at secondary sexual traits. Men, as already noted, are not only more muscular than women, they are disproportionally more muscular even when size differences are accounted for. Facial hair, too, may be relevant: beards make the male face look larger and more imposing and also permit facial expressions to be hidden. (This leaves unanswered, however, why some human groups have distinctly less facial hair than others.)

Bearded or not, young, relatively unattached, and thus reproductively unsuccessful males are analogous to those bachelor elephant seal bulls that we met in chapter 1 who have been excluded from the comparatively peaceful realm of stable, reproductive bliss by the success of the harem masters and the inaccessibility of suitable mates. Having little to lose—in a sense because they have already lost and have nowhere to go but up—they become risk-takers, often willing to engage in violence, not so much as a way of life as a last resort. A close look at homicides in Detroit during 1982 found that during a period when unemployment in the city as a whole was 11%, 41% of the perpetrators and fully 43% of the victims were themselves unemployed.[5] Examining this phenomenon, psychologists Margo Wilson and Martin Daly coined the apt phrase "young male syndrome."

Something similar occurs with crimes against property. Men aren't poorer than women, but they react differently to their poverty: they are far more likely to take things that belong to someone else. The only areas, in fact, in which women commit more crimes than men are prostitution (which some would argue is not a criminal activity but an act between consenting adults) and shoplifting, which is considerably less interpersonally aggressive than is robbery or burglary, where men predominate.

Another difference is that when women are violent, it tends to be defensively oriented, as when a woman kills a man who abuses her or her children. It is interesting that among animals as well, a mother bear with cubs, for example, is notoriously fierce, as are other females who defend their young. Thus, whereas the aggression of women tends to be reactive, men are more likely to initiate violence, to commit truly "offensive" acts.

Why aren't young, relatively unattached women similarly predisposed? Maybe they are, but if so, the predisposition is substantially reduced, in part perhaps because cultural rules typically discourage female violence. But almost certainly, underlying such expectations is the fact that female biology is differently attuned to what counts as success or failure. Otherwise, if male–female difference in this regard is simply a result of arbitrary social expectations, then we would expect that in at least some societies, things would be reversed. No such society exists. To understand why, the fundamental distinction between men and women—between sperm and eggs—cannot be overstated. A single sperm spreader can simultaneously monopolize the reproduction of many egg incubators, and in harem-keeping species, this is exactly what happens. As a result, the success of every harem master comes at the expense of failure for a number of bachelor males. How many males? If the average harem size consists of F females, then for every harem master there are on average $F - 1$ loser males (e.g., for every male with a harem of 10 females, that means that 9 males get no mates at all).

On the other hand, it is very unusual for females in a polygynous species to remain unmated. Nearly all end up in the harem of one male or another. This reduces the pressure for females to be violent or even aggressive because such risky behavior will not likely bring any particular reproductive benefit. It typically isn't all that difficult to get their eggs fertilized, especially since there usually aren't very many eggs available at any one time, and those sperm-making males not only produce huge numbers of little pollywogs, but are downright eager to spread them around. As noted, this eagerness is so great that it often results in males coming into aggressive, even violent competition with each other.

Chimpanzees and bonobos, despite their close phylogenetic relationship to human beings, are actually very inappropriate models when it comes to our "natural" mating systems. Both species live in complex, shifting, and somewhat amorphous social groups made up of multiple males and females, a system known as "polygynandry," the word itself reflecting simultaneous "polygyny" and "polyandry." If you've never heard of polygynandry before, this isn't surprising because it is a very unusual mating system, found in only a tiny number of vertebrate species and in no other primates.

As clear as it is that we aren't "naturally" monogamous, there is simply no evidence whatever that we were anything approaching promiscuous. "Humans would probably have lived . . . as polygamists or temporarily as monogamists," wrote Darwin in *The Descent of Man and Selection in Relation to Sex*:

> They would, no doubt, have defended their females to the best of their power from enemies of all kinds, and would probably have hunted for their subsistence, as well as for that of their offspring. The most powerful and able males would have succeeded best in the struggle for life and in obtaining attractive females.

Full-blown human sexual promiscuity exists only in the overheated imagination of certain writers such as the authors of the atrocious *Sex at Dawn*[6] (about which there will be more in chapter 9); followers of Sigmund Freud's highly speculative and scientifically inaccurate account of the "primal horde"; or anthropologists influenced by the work of 19th century student of native American cultures, Lewis Henry Morgan, who misunderstood the nature of Iroquois social systems, mistaking matrilineality (tracing descent through the maternal line) for promiscuity. Nothing like promiscuity has ever been reported as a standard mating system for human beings.[7]

It is the minority circumstance, but nonetheless not all that unusual, for a human society to be "matrilocal," which means that the husband is expected to move in with his bride and her family. The result is a situation in which wives are socially supported more than their husbands; and not surprisingly, women in matrilocal societies are more prone to reveal their polyandrous inclinations: more likely to engage in extramarital affairs as well as less liable to be punished for doing so. In addition, in such circumstances, divorce is generally easier to obtain. Roughly one-quarter of societies studied by anthropologists have been matrilocal, which means that about three-quarters are patrilocal, which conduces to patriarchy and the indulgence of polygyny (whether socially approved or merely de facto) over polyandry. This is because in such cases, social life occurs within a network of the husband's genetic relatives.

Chimps can be quite aggressive, even murderous, toward males from other groups; and although females frequently mate with many different males, there are also times when one male and one female leave the immediate group and form a "consortship" during which exclusive mating is the rule. Testicles have a tale to tell; in both chimps and bonobos, they are enormous relative to adult body size, strongly indicating that within a group, male–male competition is intense, although it occurs largely within the reproductive tract of each female.

By contrast, among gorillas, male–male competition occurs at the level of bodies rather than sperm. Dominant harem-holding "silverback" males have achieved their status by physically defeating and intimidating other males. They don't need to produce large quantities of sperm, because no other male will likely copulate with their females; hence, for all their great size and strength, male gorillas have comparatively tiny testicles. Human testis size, relative to body weight, is closer to that of gorillas than to either chimps or bonobos. This further cements the case that human mating as well as male–male competition occurs very differently than in the case of either chimps or bonobos, and more nearly approximates gorilla-like polygyny.

To repeat an important point made earlier: there are no human societies—either current or historically recorded—in which mating is anything like the polygynandrous chimpanzee or bonobo free-for-all. According to Frans de Waal, our foremost expert on chimpanzee and bonobo behavior

> From a female point of view, chimpanzee society seems a rather stressful arrangement. Male chimpanzees do share food with females and much of the time are on good terms with them; but males are supremely dominant, and instead of helping out with offspring, they sometimes pose a threat. Male chimpanzees hunt together, engage in fights over territory, and enjoy a half-amicable, half-competitive camaraderie. . . . With respect to male bonding and politics, chimpanzees may have the most humanlike social organization of all the apes. Nevertheless, they remain far removed from us when it comes to nurturing the young, which is entirely a female affair.
>
> Bonobo society, too, shares important characteristics with ours, such as female bonding and the nonreproductive functions of sex. The relation between males and females seems more relaxed than in the chimpanzee, yet it is evident that stable family arrangements involving both sexes do not apply to bonobos either. . . . It is no accident that people everywhere fall in love, are sexually jealous, know shame, seek privacy for sexual intercourse, look for father figures in addition to mother figures, and value stable partnerships. . . . Our species has been adapted for millions of years to a social order revolving

around reproductive units . . . for which *no parallel exists in either Pan species.*[8] [emphasis mine]

The key difference between the mating systems of chimps and bonobos on one hand and human beings on the other may well derive from differing relationships with offspring (more on this in chapter 4). As de Waal suggests

The chimpanzee male has a tendency to target infants who are not his so as to increase his own reproduction by eliminating them. The female counterstrategy is to confuse the issue of paternity: if all males are potential fathers, none of them has a reason to harm newborns. In the bonobo, females apply this strategy with perhaps even greater sophistication and combine it with collective dominance over males, a double strategy that may have been successful in eliminating this curse [infanticide] altogether.

In any event, it is obvious that chimps, bonobos, and human beings represent very different systems, driven perhaps by different evolutionary strategies. Bonobo females have evolved social behavior in which male–male sexual rivalry is unusually low key (except via the production of sperm) because there is little point in males fighting over a given female when there are numerous other receptive females available. Chimp males, by contrast, are more competitive, as well as more cooperative with regard to hunting and group territorial defense; but their society is rife with what de Waal aptly calls "the curse of infanticide" (more on this later in this chapter).

It is very common for male chimps—although not bonobos—to engage in vigorous and sometimes even lethal fights, often involving the establishment of coalitions among males and occasionally devolving into one-on-one battles when breeding opportunities are at issue. Female apes and monkeys typically organize themselves into dominance hierarchies, but only rarely with the violent consequences found among males. Dominance hierarchies, although widely considered to be aggressive, actually have the paradoxical effect of diminishing overt competition by establishing social rules of "who dominates whom." The cost to males of violating their social hierarchy is generally much greater than it is for similarly disruptive females. Among most nonhuman primates, males who attempt to mate with another male's female risk being attacked and sometimes killed. Again, it isn't surprising that the cost to males of such "acting out" is liable to be high because any success will incur corresponding cost to at least one other male whose reproductive prospects are thereby diminished and who will therefore often go to great lengths to prevent it from happening. An egg can only be fertilized by one sperm.

Of the many indications of polygyny that all people carry, the greater male penchant for violence is perhaps the most dramatic and is reflected in many ways: males of every polygynous species are more likely to initiate aggression, more likely to escalate that aggression, and more likely to win than are females or younger males. (For an informative and recent review of the fighting apparatus and

tactics of animals, see Douglas Emlen's *Animal Weapons: The Evolution of Battle.*[9])
Among primates, males are especially vigorous when it comes to guarding their
mates, whereas female monkeys and apes don't do anything comparable. In addition,
males are particularly aroused when and if they detect a potential sexual interloper—
if that interloper is male. In short, male sexual jealousy is generally intense, often
manifesting in violence, and for reasons that should by now be clear.

By contrast, a newly arrived female can be expected to be resisted by resident
females within an existing harem, just as we can expect females to be less than enthusiastic if the dominant male copulates with another female, if only because this could
result in additional youngsters who would likely compete with their own. But females
of any species that reproduces via internal fertilization can at least be confident that
their babies are in fact theirs, no matter how much sexual gallivanting is done by
the males. Female gallivanting, however—whether this involves actively going about
seeking "extra-pair copulations" or simply being receptive to an intruding male—
constitutes a direct threat to the evolutionary success of her socially identified mate.
To translate this into the language of evolutionary genetics, genes that predisposed
males to tolerate newcomer males would almost certainly have left fewer representatives over time than those that generated intolerance.

A similar situation holds for intolerance of a mate's infidelity, and once again, it
makes evolutionary sense that although a husband's infidelity may generate acute
distress in the "scorned" wife, it appears to be a pan-human, cross-cultural universal
that a wife's infidelity is far more likely to provoke not only emotional pain but often
violence directed toward the wife as well as toward her lover. Indeed, knowledge or
suspicion of infidelity is the most common identified causal factor when it comes to
wife battering and spousal homicide.[10]

Until the 1970s, it was entirely legal in several states for a husband to kill his wife's
lover if he caught them in flagrante.[11] "Homicide is justifiable," according to a Texas
statute that wasn't repealed until 1974, "when committed by the husband upon one
taken in the act of adultery with the wife, provided the killing take place before the
parties to the act have separated." In short, don't mess around in Texas! It was, however, illegal for a husband to set up a sexual liaison between his wife and another man
with the intent of "discovering" them together and then murdering his wife's paramour; the Texas law went on to note that "Such circumstance cannot justify a homicide where it appears that there has been, on the part of the husband, any connivance
in or assent to the adulterous connection." The fact that such homicide was otherwise
considered legally justifiable is powerful recognition of the violence to be expected
when a man's sexual "rights" are violated. It isn't much of a stretch to acknowledge
that in this context, sexual rights are essentially proxies for reproductive rights.

A Texas case early in the 20th century went further and permitted a cuckolded
man to kill his unfaithful wife as well, although this interpretation was subsequently
reversed:

> On July 5, 1924, Jesse Billings walked into his front room to find his wife Dolly
> and Holly Hollifield engaging in sex. Without saying a word, Billings took an

axe and "beat out the brains of Hollifield on the spot." He then caught up with his wife and beat her to death as well. Billings, convicted of murdering his wife [the killing of Hollifield was judged legal], argued on appeal that under the penal code, he was justified in killing both Hollifield and his wife. Although there were some earlier cases to the contrary, in *Billings v. State* (1925), the court rejected Billings' argument. The language of the penal code was specific, the court reasoned, it did not imply a husband had the right to kill his wife when he caught her in the act of adultery.[12]

Interestingly, Texas courts also refused to grant immunity to a wife who might kill her husband's mistress. A similar law existed in Georgia until 1977, although it went further and permitted a wife to kill her husband's paramour as well. It is worth noting that although the Texas law was structured as a legitimation of male outrage in the heat of the moment, the Georgia statue legalized killing a sexual interloper as a special case of self-defense, which also granted a father the legal right to kill his daughter's lover. (It might more accurately have been described as "gene defense.") The sexual interloper, incidentally, was not granted the right to use lethal means to defend himself.

In one Georgia case, a cuckolded husband—one J. O. Sensobaugh—who surprised his wife and lover in the act would have been acquitted had he killed the other man. Instead, Mr. Sensobaugh held him at gunpoint, saying "I don't want to kill you," tied him up, and amputated the interloper's penis with a straight-edge razor. The Court of Criminal Appeals upheld the husband's conviction for aggravated assault as well as a penalty of $300 plus 30 days in the county jail.

Until 1973, New Mexico and Utah also permitted lethal action—again, toward the other man—by a husband who caught his wife in the act of adultery, although in the case of New Mexico, the act was seen as biologically excusable rather than socially admirable:

"The purpose of the law," wrote a New Mexico court,

> is not vindictive. It is humane. It recognizes the ungovernable passion which possesses a man when immediately confronted with his wife's dishonor. It merely says the man who takes life under those circumstances is not to be punished; not because he has performed a meritorious deed; but because he has acted naturally and humanly.

Much legal attention had been given to what might seem a prurient detail: ascertaining how precisely must the husband know that his wife was actively engaged in adultery at the moment he discovered her with another man. In one case, hearing squeaking bed springs was deemed sufficient; in another, finding a wife in her bedclothes along with a man with his shoes off.[13] Before it was eventually deleted from the penal code in 1974, the Texas law legalizing a husband's killing of his wife's lover led to some interesting permutations. For example, in 1971, David Smith surprised William Sumner having sex with Smith's wife, whereupon Mr. Smith shot and killed Mr. Sumner, whose wife then attempted to collect on Sumner's $5,000 "accidental death" life insurance policy. The insurance company denied the claim, maintaining that having committed adultery, Sumner could not be considered to have died accidentally; rather, his death was tantamount to suicide—in Texas, at least. An appeals court reversed this action, maintaining that even though Texas law permitted homicide by an "outraged" husband, an adulterer should nonetheless not anticipate death as a likely outcome of his action: "ordinarily most husbands do not vindicate such wrongs by homicide but lay their problems in the lap of divorce court," an affirmation, perhaps, of the rule of law over the promptings of biology.

Yet another Texas case—this one in 2007—revolved around the killing of a man, Devin LaSalle, by Darrell Roberson, husband of Tracy Roberson, after Mr. Roberson came home unexpectedly and found his wife in sexual congress with Mr. LaSalle in the latter's pickup truck, parked in the driveway. It appears that upon seeing her husband, Tracy Roberson called out that she was being raped, whereas in fact, she and LaSalle were lovers, as evidenced by an earlier text message she had sent to him saying "Hi friend, come see me please! I need to feel your warm embrace! If ur unable to I completely understand!!! Call me."[14]

Mrs. Roberson was subsequently convicted of manslaughter and given a five-year prison term for having induced a killing based on a false accusation. This led to such notable headlines as "Husband Allegedly Kills Wife's Lover; Wife Indicted." The jury concluded that when Darrell Roberson killed Devin LaSalle, he thought that he was saving his wife from rape (or perhaps exacting retribution for the act), whereas he was actually killing his wife's lover, after his wife, thinking quickly and doubtless desperately, cried "rape" to absolve herself—and presumably without reckoning on her husband's lethal response.

Historically, in many cultures, the murder of an adulterous wife and/or her lover has not only been condoned but encouraged. On the island of Yap (one of the Caroline Islands in the Western Pacific), a cuckolded man "had the right to kill [his wife] and the adulterer or to burn them in the house." Among a tribe known as the Toba-Batak of Sumatra, "The injured husband had the right to kill the man caught in adultery as he would kill a pig in a rice-field." The Nuer people of East Africa recognize that "a man caught in adultery runs a risk of serious injury or even death at the hands of the woman's husband." Among the Melanesians from the island of Wogeo, every-one understands—even expects—what a local informant describes as "the rage of the husband who has been wronged. . . . He is like a man whose pig has been stolen," but even angrier. These are not isolated, unusual cases; they are found wherever human beings abide.[15]

The bottom line: male lethality is unquestionably real, especially in the context of sexual competition, and even more when the competition is immediate and involves direct access to a man's sexual "rights" and "property," which is to say, women. An interesting question is whether such violence is more likely in the context of polyg-yny than monogamy. In a monogamous species, a male's ultimate reproductive suc-cess is totally dependent on his mate's; so if monogamy is 100%, there is no difference between the two. As a result, we might expect monogamous males to be, if anything, *more* intolerant than their polygynous cousins of their mates' sexual infidelity, and also more violently disposed toward their mates' lovers.

It isn't currently known whether polygyny exaggerates male sexual jealousy, although there is some evidence that it might. Thus, we've already noted that one effect of polygyny is to generate sexual dimorphism, with males selected to be large and ferocious, mostly because such individuals are more successful in the male–male competition that results in some males emerging as harem masters. As a result, with harem-holding males already likely to be evolutionarily outfitted with physical weapons, as well as predisposed to greater aggressiveness and a low threshold for violent conflict, it is probable that they are especially preadapted to respond vigorously when their sexual monopoly over "their" females is threatened.

There is a well-established history of harem-holding men using all sorts of stringent means to defend their bevy of wives. When the harem is large, which only occurs when the harem holder is especially powerful (socially at least as much as physically), the women have always been closely guarded, often by eunuchs. Trespassing on such a harem was perilous indeed, with violators risking death, not uncommonly after first being castrated for good measure. It may be that polygynous men are especially jealous of their wives, just as wealthy people are reputed to be particularly anxious about losing their wealth. "De folks wid plenty o' plenty," we learn in the song "I Got Plenty of Nuthin'" from *Porgy and Bess*, "Got a lock on de door, 'Fraid somebody's a-goin' to rob 'em, While dey's out a-makin' more."

In evolution's calculus, "more" translates to more children. And the robbery? Stealing another's genetic patrimony.

In any event, whether or not polygyny gives rise to more male violence than does monogamy, there is no question that violent mate guarding only occurs in a world in which women—at least, some women—are going to engage in sex with other males. Social taboos don't prohibit behavior that people wouldn't otherwise do. Just as there are no formal rules that prohibit eating dog poop, because no one is inclined to do so, there are numerous civil laws prohibiting murder, theft, coveting one's neighbors wife, and so forth because "human nature" is such that sometimes people are inclined to do just these things. If women weren't polyandrous—selected to engage, on occasion, in sexual relations with more than one male—then men (whether polygynous or not) wouldn't be disposed to violent mate guarding in the first place.

E ven when not mate guarding, male aggression and violence can readily confer a reproductive benefit on the perpetrator. University of Texas evolutionary psychologist David Buss listed the following potential payoffs to especially aggressive or violent individuals: "expropriating resources, defending against incursions, establishing encroachment-deterring reputations, inflicting costs on rivals, ascending dominance hierarchies, dissuading partner defection, eliminating fitness draining offspring, and obtaining new mates."[16]

Significantly, the likelihood that a man will kill another man is much higher in societies with large disparities in wealth, situations in which men are particularly pressed to achieve the status needed to acquire and maintain relationships with women. It could be argued that the cause of heightened male–male homicide can be found in poverty rather than income inequality itself, because poorer societies also tend to be more unequal in their wealth distribution. In this regard, Canada provides an interesting semi-experimental test because its Atlantic provinces are significantly poorer than those in the west, and yet their social welfare policies are more generous, resulting in less income disparity. It turns out that inequality predicts homicide rates, whereas total levels of wealth or poverty does not.[17]

Polygyny, along with the reproductive inequality it involves, occurs only in a context of socioeconomic inequality. It is both a manifestation of inequality and a generator of it because it unbalances the sex ratio among marriageable adults. As anthropologist Napoleon Chagnon[18] has emphasized, a man who has 10 times as many children and women as is his "due" must either work 10 times as hard to support, maintain, and defend them, or he must take what he needs from other men.

The various payoffs to male violence—from "resource acquisition" to status enhancement—translate ultimately into evolutionary fitness, which is success in projecting genes into the future. The most obvious are those directly connected with mating success. Aside from deterring sexual competitors and thus making one's own paternity more likely, aggressors can benefit by getting the stuff—technically, the "resources"—of others, especially other men. Insofar as women have a choice of husbands, men who possess resources (in short, wealth) are more attractive to women as well as to women's families, who, if we look at human behavior worldwide, often sell their daughters in return for a "bride price"—essentially the inverse of a dowry (more on this in chapter 4).

In Islamic law, a man is entitled to up to four wives, the number strictly determined by his income. In traditional societies, there is a clear correlation between men's status—itself often achieved by violence or the threat of violence—and the number of his wives. Saudi Arabia's King Abdullah, who died in January 2015, never had more than four wives . . . at a time. Altogether, he was married 13 times and produced at least 30 children; the exact number is subject to debate. And Abdullah was ostensibly a "modernizer."

According to Laura Betzig, who analyzed data from 104 different human societies, the size of men's harems—especially characteristic of historically despotic societies—was directly predicted by their "hierarchical power," which in turn she succinctly described as "the exercised right . . . to murder their subjects arbitrarily and with impunity."[19]

Minor kings would typically have a harem of about 100; kings of greater substance, perhaps 1,000; and emperors, 5,000 or more. A similar pattern occurs worldwide, even as the ecological details vary. In a recent overview of the subject, Betzig concluded that as calorie intake increases with more sedentary lifestyles, from hunter-gatherer, to herder-gardener, to full-time farmer societies, the range and variance of women's reproductive success doesn't change, whereas that of men increases significantly.[20]

Within each society, there are substantial differences in reproductive success—sometimes called "reproductive skew"—notably among men. Among the Aché hunter-gatherers of Paraguay, more lethal hunters have more sexual partners and more offspring. On the Micronesian atoll of Ifaluk, men's wealth and status is attained by physical abilities—notably the capacity to intimidate and otherwise outcompete other men—and this, in turn, correlates closely with reproductive success among them. The same can be said of the Kipsigis people of modern-day Kenya (where wealth is measured in either number of cows or acres of land), 19th-century Mormons in the United States, and in 15th- and 16th-century Portuguese.

Anthropologist Napoleon Chagnon has reported that among the Yanomamo of the Venezuelan Amazon, men who have killed other men (so-called *unokais*) have more wives and more offspring than do other men who have not killed. Although this finding has been disputed, and granted the importance of distinguishing between personal violence and a "war instinct," it is clear that the Yanomamo, who call themselves the "fierce people," do so for good reason.

This ferocity, notes Chagnon, is nearly always provoked by something related to sexual access to women: "Most fights begin," he writes

over sexual issues: infidelity and suspicion of infidelity, attempts to seduce another man's wife, sexual jealousy, forcible appropriation of women from visiting groups, failure to give a promised girl in marriage, and (rarely) rape. . . . The most common explanation given for raids (warfare) is revenge for a previous killing, the most common explanation for the initial cause of the fighting is "women."[21]

When Professor Chagnon explained to his subjects that some anthropologists believe that when the Yanomamo fight and occasionally make "war," they do so over animal protein, he was told, with a chuckle, that indeed, they like meat—but they like women more.

In his book, *Eve's Seed: Biology, the Sexes and the Course of History*, historian Robert McElvaine[22] provides an eye-opening review of the exploits—sexual no less than geopolitical—of various modern-day, mostly American and European, male political and military leaders, suggesting that much of the excessive amatory exploits of JFK, for example, derived from a need to prove their manliness. I agree, but add that they weren't simply acting out via a socially constructed, symbol-rich behavior; they were also acting out a biological . . . what? Not an imperative, but a predisposition, a penchant, an inclination, a tendency, scratching an itch that is to some extent evolution's bequeathal to all men. Problematic in itself, this male itch scratching is more troublesome yet when it manifests itself in orchestrated violence and conquest on a large scale.[23]

For female mammals, reproductive success is mostly limited (which is to say, "determined") by access to resources needed for the production and rearing of children. For males, females themselves comprise the key limiting resource. As we have seen, this is why male–male competition is so prominent, and so prominently aggressive, whereas female–female competition, although real enough, tends to be more subtle and typically focused more on investing effectively in their offspring. Males, by contrast, are more concerned with creating the circumstances necessary for fathering these offspring in the first place.

Beyond this, male aggression can benefit the aggressor by enhancing his position in the social hierarchy, which in turn often translates into mating success; among nonhuman primates (as with mammals generally), higher-ranking males nearly always have greater mating access to females, especially when the latter are ovulating. And the data are clear that among human beings, dominant and otherwise successful men have more sexual partners and, at least in most traditional societies for which data exist, more offspring.[24]

According to the *Guinness Book of World Records*, the prize (or at least the maximum number) for the most children borne by a woman belongs to a 19th-century Russian woman who had 69 live births in the course of 27 pregnancies, each of which produced twins, triplets, or quadruplets. Most women, of course, have many fewer! But 69 offspring for one woman—astounding as it is—is dwarfed by the number of children fathered by exceptionally "fit" men. Although nothing is known about the violent or nonviolent propensities of this remarkably fertile woman, we know quite a lot about the inclinations of hyper-reproductive men, not only that their offspring exceed in numbers those of the most prolific women by an order of magnitude (recall that males are the high-variance sex), but—not surprisingly—that cases of extravagant male reproductive excess go along with extravagant violence. (Note once again, however, that such reproductive performance necessitates reproductive *failure* for a corresponding number of other men in the same society.)

For a long time, one of the most renowned examples of polygyny, violence, and reproductive performance converging in a single individual was Moulay Ismail, legendary ruler of Morocco, said to have fathered 888 children and someone whose eagerness to maim, torture, and murder his opponents (and even many of his supposed allies on occasion) evidently justified his name, Ismail the Bloodthirsty.

It has recently been found, however, that Ismail was by no means the most reproductively notable man. It turns out that fully one out of every 200 men alive today worldwide is a direct descendant of Genghis Khan; that's roughly 16 million men.[25] This was determined by assessing the abundance of the Great Khan's Y chromosome, so it considers only male descendants. It seems reasonable to assume that Genghis gave rise to no fewer female descendants, which would increase his paternity to something like 32 million people. Not surprisingly, Genghis Khan gave us this short version of his personal philosophy and approach to a fulfilled life: "Happiness lies in conquering your enemies, in driving them in front of you, in taking their property, in savoring their despair, in raping their wives and daughters."[26]

The connection between violence and polygyny is notably tight, just like the close coupling between maleness and violence. There are at least two processes whereby violence predisposes toward polygyny. For one, men are likely to be successful harem keepers in proportion to how violent and intimidating they are. And for another, societies that practice institutionalized violence—especially those engaging in chronic warfare—are prone to developing a shortage of men, which not only reduces the socially disruptive effects of polygyny (because in such societies there are fewer unmated males to make trouble), but might also predispose toward polygyny simply because of the resulting excess of unmated women, even if monogamy is otherwise encouraged. In his *Principles of Sociology*, the 19th-century social philosopher Herbert Spencer noted that polygyny is more common among societies that lose many male soldiers in war; he proposed that polygyny occurs as a means whereby these societies maintain their population. Modern biologists would disagree, suggesting that it is a means whereby some men take advantage of social conditions to maximize their own reproduction.

In some especially obnoxious cases—especially those involving child "brides" who are essentially dragooned into polygyny (which is to say, raped)—young women also have little or no choice about their reproduction. In other cases, women seem to be active participants in their polygynous participation (more in chapter 4). After all, despite the fact that women are physically less intimidating than men and often less socially powerful as well, they, no less than men, have been selected to maximize their own Darwinian fitness insofar as options exist. This assumption doesn't require, incidentally, that women or men are consciously aware of doing so. So it isn't surprising if, in a society with a demographic shortage of men, many women willingly comply with polygyny. Being the nth wife of a successful man is likely to be more appealing than being either a single mother or nonreproductive. And it isn't necessary to distinguish whether this "decision" is urged by social pressure—compounded by fear—or by evolutionary predilections, whispering deep within, insofar as the two push in the same direction.

In any event, Genghis Khan may well have been extraordinary, as well as extreme, but he certainly wasn't unique in leaving more than his share of descendants. A fellow known as Giocangga, the great-granddaddy of what eventually became the Qing dynasty of China, gave rise to roughly 1.6 million men who are alive today.[27] If Giocangga had been an ordinary man, he would have perhaps 100 living descendants in the modern world; so his genetic impact—his fitness payoff compared with the average man—is 80,000 times more than expected if he wasn't an unusually successful representative of the high-variance sex.

The list can be multiplied ad nauseam, but here are just a few more. King Solomon is said to have had no fewer than 700 wives (we don't know how many children). In the 20th century, Sobhuza II, king of Swaziland for more than 40 years, had 70 different wives with whom he fathered no fewer than 210 offspring. Jacob Zuma, president of South Africa, has been married five times, is currently and simultaneously married to three women, and is believed to have fathered 22 children. Elizabeth Taylor was married nine times. The renowned, now deceased, Nigerian musician Fela Kuti was wed to 27 women in one all-embracing ceremony; his renown derived not from this, but from his musicianship, which doubtless contributed mightily to his polygyny. (Violence isn't the only way that harem masters achieve their status, but it is certainly a prominent route. Indeed, social prominence itself can be a key to polygyny; see chapter 7 where I examine the possible connection between polygyny and "genius.")

Among many traditional societies, the leaders are often called "Big Men"; and indeed, these people, who not uncommonly have big harems as well, tend to be literally big men. People aren't elephant seals; if nothing else, we are less polygynous (at least most of us!). But the success of certain extreme men would make an elephant seal green with envy. Although neither Ismail or Genghis was 2.5 times the size of their wives, they nonetheless multiplied, via culture, their ferocity and violence many more times than that.

A man needn't be an Ismail or Genghis for certain male traits—many of them less than admirable in polite company—to yield an evolutionary payoff. The unfortunate truth is that just a smidgeon of assertiveness, aggressiveness, and otherwise annoying macho behavior has likely been reinforced in our evolutionary past, even if such traits aren't admired (at least not publicly) these days. It has been well established that a male–female difference in aggressiveness is apparent among children as young as 12 months of age, especially when it comes to what is known among developmental psychologists as "rough-and-tumble play."[28] Moreover, this difference is cross-cultural.[29] Thus, it cannot simply be attributed to differences in socialization and local family traditions.

Boys also generally develop more slowly than do girls. They often lag behind in cognitive development, language acquisition, and so forth. This also appears to be true of other polygynous species, which makes sense as a factor associated with sexual bimaturism (chapter 1). It can be predicted that more monogamously inclined species—such as marmosets, gibbons, and titi monkeys, among primates—would be more equal in male–female development rates. It is suggestive that among the mostly monogamous owl monkeys, male and female juveniles do in fact grow at the same rates.[30] Even among nonhuman primates—or at least, in the one species that to my knowledge has been tested—juvenile males preferred to play with a toy truck and females with a toy doll![31] As a general principle, play behavior in polygynous species is more sexually dimorphic than it is in those that are more monogamously inclined.[32] Without going into the details, a "robust" finding is that among human beings, across a diversity of cultures, there are significant and consistent differences between boys and girls when it comes to their preferred play patterns: boys are more physical and aggressive, while girls are more interested in dolls and behavior suggestive of mothering and caretaking.[33]

Those committed to the primacy of social learning theory nonetheless pointed out nearly a half-century ago that if young boys—and not girls—are positively reinforced ("rewarded") for their aggressiveness, such behavioral differences could have been shaped by their different experiences.[34] And yet a study of the aggressive behavior of Norwegian preschool children found that little boys were reprimanded *more* than girls for their aggression, and yet they remained consistently more aggressive.[35]

Male–female differences in children—even very young ones—show up in other contexts as well. For example, as early as 12 months of age, boys are more inclined to leave their mothers' sides in novel situations, and girls are more liable to be frightened by artificial masks showing scary human faces.[36] Boys don't strive to be double mother suckers, as described in chapter 1, but they do the human equivalent.

It is downright dreary to have the same basic, male chauvinist distinctions between men and women, boys and girls, confirmed in study after study, starting in very young childhood, but there's no way around it. As described by anthropologist/physician Melvin Konner

> Subjects ranging from two-year-olds to adults were studied by many investigators with techniques including direct observation, parents' ratings, teachers' ratings, laboratory experiments, personality tests, self-descriptions, and fantasy during play. For nurturance and affiliation, forty-five of fifty-two studies found females showing more; two found the reverse, and five found no difference. But for overt physical aggression, girls and women showed less in *all* studies. Reviews and meta-analyses of hundreds of studies in more than four decades strongly confirm these findings.[37]

The behaviors measured ranged from overt indications of hitting, kicking, and rock throwing to verbal threats. These male–female, boy–girl sex differences in aggressiveness persist throughout childhood and adolescence, with the difference between male chauvinist piglets and their female counterparts becoming most pronounced during young adulthood.[38]

In a now famous research project that compared the behavior of boys and girls in six different cultures—India, Japan, Kenya, Mexico, the Philippines, and the United States—developmental psychologists Beatrice and John Whiting examined such behavioral events as "seeks help," or "touches" (on the nurturance side) and "reprimands" or "assaults" (aggressive) and found that boys were more egoistic and more aggressive, whereas girls were (surprise!) more nurturant.[39] Interestingly, there were some cases in which girls from one culture were on average more aggressive than boys from another, showing the role of cultural conditioning and social learning. More interesting yet, within any one culture, the girls were consistently less aggressive and more nurturant than the boys, showing the degree to which biology pries the sexes apart, regardless of social circumstance.

The process of behavioral development is complex, with biology and social experience interdigitating so as to be almost inextricable. For example, given the opportunity, children in essentially all cultures tend to self-segregate by sex: boys mostly play with boys and girls with girls after they are no longer toddlers. Moreover, adults consistently treat boys and girls differently. Both of these tendencies enhance male–female differences that are already present.[40] So, male–female differences—beginning early in life—are real, almost certainly because they are derived from such biological factors as early exposure to hormones, which bias brain development and subsequent behavior; but they are also exaggerated and thus made even "more real" by cultural practices.

It needs to be emphasized, by the way, that boy–girl, male–female behavioral differences cannot simply be explained by differences in testosterone and not merely because testosterone levels are very low in young children of either sex. There is no simple, one-to-one relationship between testosterone levels and aggression,[41] although a broad, general correlation nonetheless exists. And second, even if androgenic ("male-stimulating") hormones in general and testosterone in particular were directly responsible for greater male aggressiveness, the evolutionary aspect of this behavioral difference would still require additional explanation of the sort attempted in this book. Let's imagine a caricature of reality: that a particular chemical "causes" a particular behavior. This explanation operates at the level of what biologists call proximate causation, the immediate precipitating factor. It doesn't explain ultimate causation, the underlying reason why evolution has generated a system that works this way in the first place. For that, we need to understand not only the "how" behind the behavior in question but also the "why."

How are internal combustion automobile engines powered? By gasoline, or diesel. Why? Because that's *how* they were designed. They could have been made otherwise: for example, to run by electricity. Gasoline or diesel is only the "proximate" fuel for most vehicles.

The question pursued in this chapter isn't so much *how* it is that boys and men are more aggressive than girls and women, whether they run by gasoline, diesel, or electricity—whatever the specific combination of hormones, brain mechanisms, and responsiveness to social experiences that causes these differences—but *why* evolution has produced this distinction; that is, what the "design" considerations are, regardless of how they are achieved. There is no inherent characteristic of testosterone that requires it to promote aggressiveness; it could just as well predispose a living thing to pacifism, if that was evolution's goal.

There is also no doubt that experience matters too and that male–female differences in experience also occur. It is entirely possible that boys begin life being slightly more aggressive than girls, that they are in turn treated differently as a result, and that this difference is subsequently magnified over time. And whereas it could be merely coincidental that such differences correspond with the differences that natural selection would otherwise generate, it is more parsimonious to assume that a genuine connection exists. It is also conceivable that parents just happen to treat boys and girls differently in a manner that produces greater male aggressiveness; that such a difference contributes positively, on average, to male biological success; and that this series of coincidences just happens to occur cross-culturally. But it is unlikely.

It is surprising how persistent certain ideas can be, even when proven wrong. For example, Margaret Mead made the general case in 1949[42] (expanding her prior descriptions of specific tribal societies) that male–female differences in behavior resulted solely from different cultural traditions, which themselves differed at random from society to society. To a remarkable extent, this remains the conventional wisdom in anthropology, psychology, and sociology today, despite the fact that cross-cultural evidence has consistently shown not only that boys and girls are treated differently but that these differences reliably emphasize different traits depending on the gender of the children. Boys are socialized in the direction of sex stereotypes that emphasize such behaviors as "active, adventurous, ambitious, arrogant, and assertive" (and that's only the "a's"), while girls are socialized to be "gentle, kind, meek, mild, pleasant, and sensitive."[43]

Whether tested in Canada, Peru, Pakistan, Nigeria, or Norway, it turns out that sex role stereotypes are remarkably similar. Once again, it could be coincidental that across different societies, young children are socialized into precisely those behavioral roles that are consistent with the underlying biological substrate that distinguishes males from females. More likely, such socialization has been independently derived by these societies because the outcome—different behavioral tendencies between men and women—has proven to be compatible with children's biological inclinations as well as with perceived favorable outcomes (social as well as biological) later in life.

This may surprise you. There is a single gene, clearly identified and known as SRY, that dooms its carriers to shorter lifespans, greater probability of death due to "accidents," as well as increased risk of being not only violent but also the victim of violence. The impact of this gene is such that substantially more than one-half of people who have run afoul of the law and are currently incarcerated carry it—although to be fair, it should be noted that nearly one-half of nonfelons are similarly afflicted. Good luck to those unfortunates forced to deal with such a heavy genetic liability. Although we know a lot about this particular woe begotten load of DNA, there is no reliable medical or psychological cure available for anyone burdened with it. The SRY gene is located on the Y chromosome, and if you haven't already guessed, it's the one that makes its carrier male (SRY comes from "Sex determining Region of the Y chromosome").[44]

Even though many, perhaps most societies are sexist in that they give social, economic, and political preference to men over women, and boys over girls, and even though women face an additional mortality factor in childbirth, it is almost universally true that men's lives are shorter than women's.

One possible explanation is that because men are chromosomally XY and women are XX, and because the Y chromosome is notably depauperate when it comes to genetic material, perhaps women have a lifespan advantage because they simply have a deeper genetic reservoir. (We might accordingly say that men suffer from "X chromosome deficiency disorder.") But this doesn't explain why the same male–female mortality difference is found among birds as among mammals, although in the avian world females are the depauperate "heterochromosomal" sex. A consistent pattern found in birds and mammals is that males die younger. This is notably the case for polygynous species generally, whereas among monogamous ones, males and females face similar mortality schedules.[45]

In that species known as *Homo sapiens*, and in every society for which data are available, young adult and adolescent males are more at risk from accidents (often associated with showing off).[46] Another cross-cultural universal is that although both girls and boys typically undergo initiation rituals, these rites of passage are consistently more severe for the sperm makers. Compare debutante balls or Latin American *quinceañera* events with mandatory vision quests, killing a lion, and so forth. Many traditional societies—perhaps most—require that as part of menarche (first menstruation), girls are either isolated in specified ways or otherwise identified as having achieved womanhood, but it is rare for this transition to be especially painful.

Female circumcision is a notable apparent exception, but as barbaric as this practice is, it nearly always takes place when the girls are quite young and thus isn't an initiation into adulthood as such. Moreover, it seems clear that female circumcision—a horrifying event that ranges from clitorectomy to literally sewing up the vagina—is based on male control of female reproduction, rather than female social initiation comparable to the male experience. (More on this phenomenon in chapter 6.)

It isn't obvious whether vigorous and often violent male initiation practices are a consequence of the greater violence in male lives or if such trials predispose young men toward greater violence; but either way, the connection is undeniable.

Moreover, it seems likely that much of the harshness of male sexual initiation occurs as a consequence of male–male competition itself, with older men subjecting younger ones to socially prescribed trials and tribulation. A good review of this situation was titled "Male Genital Mutilation: An Adaptation to Sexual Conflict."[47] There is little doubt that higher male mortality, as well as greater male propensity for violence, derives not from any particular gene but from a deeper, evolution-based liability: maleness itself (genetically keyed to SRY) and its nearly unavoidable connection to polygyny and all that it entails. We've already noted some of the early developmental distinctions between boys and girls, and to some degree, as the twig is bent, so grows the branch, even the whole tree. At the same time, being male opens the door to an array of circumstances that evolutionary biologists are only now beginning to understand.

A study of male–male sexual assaults in prisons concluded that the typical man who rapes another man "does not consider himself to be a homosexual, or even to have engaged in a homosexual act. This seems to be based upon his startlingly primitive view of sexual relationships, one that defines as male whichever partner is aggressive and as homosexual whichever partner is passive."[48] It appears that the motivation in such cases is at least as much violent, aggressive, and dominance seeking as it is sexual. And insofar as it is sexual, too, that italicizes the regrettable connection that often exists between maleness, violence, and sex.

Let's look a bit more closely at male initiated lethal violence. Start with wars. No one seriously doubts that men are the primary war makers, as well as the predominant war fighters. A review of 75 different human societies that engaged in warfare found that conflict over women was the most cited cause: in 34 cases. And conflict over resources was cited in 29 of the remaining 41 cases; it is noteworthy that resources often serve as a proxy for access to women and thus reproductive success (by providing opportunities for paying bride price or otherwise enabling the acquisition of additional wives and/or matings).[49] To be sure, there are women warriors, especially these days, as many countries struggle to provide "equal opportunity" regardless of sex. But both historically and worldwide, men—notably, young men—have been overwhelmingly the ones who wage war, and—no coincidence here—it has typically been old men who encourage them to do so (possible implications are considered in chapter 5).

Wars are bad enough, and when it comes to civilian casualties, women and children die along with men. Leaving this aside, what about plain old day-to-day homicides? In late 2014, a United Nations report concluded that

> Homicide and acts of personal violence kill more people than wars and are the third-leading cause of death among men aged 15 to 44. Around the world, there were about 475,000 homicide deaths in 2012 and about six million since 2000, making homicide a more frequent cause of death than all wars combined in this period.[50]

In their magisterial treatise *Homicide*,[51] psychologists Martin Daly and Margo Wilson presented a raft of data on the human tendency to kill other human beings. Among their findings, not only are men—especially young men (i.e., those at an age when they are seeking to establish their reproductive situation)—overwhelmingly the perpetrators, but they are also overwhelmingly the victims, too. More than 95% of homicides involve young men killing other young men, a pattern that held over a great swath of human history—in fact, wherever suitable data exist and for whatever time periods—and also cross-nationally in modern times.

After reviewing murder records over a wide historical range, and from around the world, Daly and Wilson concluded that "The difference between the sexes is immense, and it is universal. There is no known human society in which the level of lethal violence among women even begins to approach that among men."

Daly and Wilson also found that a man is about 20 times more likely to be killed by another man than a woman is by another woman. This holds true for societies as different from each other as modern-day urban America (Philadelphia, Detroit, and Chicago), rural Brazil, traditional village India, Zaire, and Uganda. This is not to say that actual murder rates are the same in these different places. In modern Iceland, for example, 0.5 homicides occur per million people per year; whereas in most of Europe, the figure is closer to 10 murders per million per year; and in the United States, it is over 100. If anything, these results *underestimate* the preponderance of male–male competition. Thus, many of the female–female homicides reported in large-scale samples are actually infanticide. When Daly and Wilson reworked these results to include only those cases in which the killer and victim differed in age by 10 years or less, the male–female difference was revealed to be even more impressive.

The crucial point for our purpose is that despite wide differences between countries, the basic male–female pattern remains stable: male–male homicide exceeds its female–female counterpart by a whopping margin. In addition, the fact that the ratio of male–male to female–female violence remains remarkably *un*varying from place to place argues for its biological underpinnings and parallels the male–male competition seen in other moderately polygynous species.

The same trend can be found across history. Thus, even though a 13th-century Englishman was 20 times more likely to be murdered than an Englishman is today, he was 20 times more likely to have been murdered by another *man* than an Englishwoman was by another woman. Not only that, but around the world and across time, the age of the vast majority of these male murderers remains constant, in their mid-20s.

Although in recent years women have been increasingly involved in criminal behavior, Daly and Wilson cite FBI statistics attributing this increase to growing numbers of women arrested for "larceny-theft," whereas the proportion of women arrested for violent crimes—and for homicide in particular—has declined slightly.

In 1958, sociologist Marvin Wolfgang[52] published what has long been the classic study of homicide in America based on nearly 600 murders in the city of Philadelphia. Trying to explain why more than 95% of the killers were men, Wolfgang—like so many sociologists a proponent of learning theory and cultural explanations—wrote, "In our culture [the average female is] ... less given to or expected to engage in physical violence than the male." We are supposed to infer that things are different in other cultures, but this simply is not so.

Writer and traveler Peter Matthiessen reported that among the Dani people of the New Guinea highlands, for example

> A man without valor is *kepu*—a worthless man, a man-who-has-not-killed. The *kepu* men go to the war field with the rest, but they remain well to the rear.... Unless they have strong friends or family, any wives or pigs they may obtain will be taken from them by other men, in the confidence that they will not resist; few *kepu* men have more than a single wife, and many of them have none.[53]

Manuel Sanchez, a 32-year-old man from Mexico City, summed up the situation there:

> Mexicans, and I think everyone in the world, admire the person "with balls," as we say. The character who throws punches and kicks, without stopping to think, is the one who comes out on top. The one who has guts enough to stand up against an older, stronger guy, is more respected. If someone shouts, you've got to shout louder. If any so-and-so comes to me and says, "Fuck your mother," I answer, "Fuck your mother a thousand times." And if he gives one step forward and I take one step back, I lose prestige. But if I go forward too, and pile on and make a fool out of him, then the others will treat me with respect. In a fight, I would never give up or say, "Enough," even though the other was killing me. I would try to go to my death, smiling. That is what we mean by being "*macho*," by being manly.[54]

There is a powerful bias in much of the scholarly world, promoted by renowned psychologists, anthropologists, and sociologists, that male–female differences have been created almost entirely by differences in upbringing and social expectations. As a result—whether by error or preexisting bias—social scientists have contributed to a vast myth: that of the equipotential human being, the idea that everyone is equally inclined to behave in any way. Equipotentiality is an appealing sentiment, attractively egalitarian. There is only one problem: it isn't true, especially when it comes to male–female differences in violence.

Quite simply, the presumption of equipotentiality flies in the face of everything known about the biological underpinnings of behavior and of life itself. Daly and Wilson usefully coined the phrase "biophobia" to describe this belief in the interchangeability of people and the all-encompassing role of learning and social tradition. Those in the grip of biophobia continue to attribute the huge gender gap in violence to local circumstances and traditions, to social expectations and learning, as though evolution has had no part to play.

With the benefit of evolutionary insight, much that had previously seemed random or incomprehensible now makes a grim kind of sense. When Marvin Wolfgang conducted extensive interviews with convicted killers in Philadelphia, for example, he was able to identify 12 different categories of motive. Far and away the largest, accounting for fully 37% of all murders, was what he designated "altercation of relatively trivial origin; insult, curse, jostling, etc." In such cases, people got into an argument at a bar over something as seemingly unimportant as a sports game, who paid for a drink, an offhand remark, or an apparently casual insult, and so forth.

To die over something so inconsequential as a random comment or a dispute about some distant event seems the height of irony and caprice. But in a sense, disputes of this sort are not trivial, for they reflect the evolutionary past when personal altercations were the stuff on which prestige and social success—leading ultimately to biological success—were based. It is not surprising, therefore, that young men today will fight and die over who said what to whom, whose prestige has been challenged, and so forth.

Finally, I cannot do "justice" to male-initiated violence without a (mercifully brief) look at infanticide. When we think of male–male competition, it is only natural to focus on what males do to each other. And yet, male–male competition occurs in many domains, sometimes without any direct contact between the competitors. For example, men who strive to accumulate wealth (or resources generally) may do so without necessarily beating each other over the head, although the competitive component is nonetheless there, and the payoff to success is every bit as real, and no less reflected ultimately in enhanced fitness.

Such payoffs typically accrue via "precopulatory" competition in which winners are rewarded by enhanced mating opportunities. Recently, however, biologists have come to recognize that competition—especially the male–male variety—doesn't end with mating; they have accordingly begun to study "postcopulatory" competition. For example, in many species semen tends to harden after intercourse, producing a rubbery mating plug that keeps other males from displacing the sperm already in place. In many invertebrates, the male's penis resembles a Swiss army knife, with various scrapers and other devices adapted to remove a predecessor's semen. But undoubtedly the most dramatic—indeed, tragic—cases of male–male, postcopulatory competition occur following a male takeover in certain polygynous species.

The iconic example is the Indian langur monkey, studied by Sarah Hrdy for her doctoral dissertation in anthropology at Harvard University and reported in her fine book, the *Langurs of Abu*.[55] Langurs are typical polygynous primates in which ever-troublesome, excluded bachelor males periodically succeed in overthrowing the dominant harem master. Shortly thereafter, squabbling breaks out among the victors; and out of the postrevolutionary chaos, one male emerges to begin his reign as the new despotic harem holder. And this is when things really get ugly.

The newly ascendant male mates with the adult females. There is no surprise here. What is deeply disturbing, however (although fully concordant with evolution's hard-hearted, gene-based calculus), is that he also proceeds to kill the still-nursing infants. This induces their mothers to begin cycling once again, whereupon they mate with the infanticidal male. In most mammals, lactation inhibits ovulation; this "lactational amenorrhea" occurs, although to a lesser extent, among human beings too. The upshot is that by killing the infants, the new harem master increases his own reproductive output while diminishing that of his predecessor. Hence, infanticidal males are acting to further their own fitness, at the expense of the recently ousted harem master—and, of course, to the distinct disadvantage of those infants and their mothers. (For their part, the bereaved mothers are simply maximizing their fitness by mating with their infants' murderers. Human morality doesn't apply.)

When Hrdy first described the carnage, notably in her article "Infanticide as a Primate Reproductive Strategy,"[56] the traditional scholarly establishmentarians—especially within the social sciences—were either incredulous or outraged. It must be an artifice of something or other, they claimed. Maybe the animals were overcrowded, or protein starved, or disturbed by the human observer. At that time, many people believed that evolution acted "for the good of the species," making infanticide not only ethically abhorrent but bad for the species and therefore even harder to believe. But in fact, we now understand that evolution operates at the level of individual and gene benefit, with any "good of the species" simply an incidental, summed, arithmetic by-product of what happens at the lower, more functional levels. And much as we might wish otherwise, natural selection is neither moral nor immoral; rather, it is thoroughly amoral, promoting whatever works at those lower levels. And infanticide definitely works.

Infanticide is such a repulsive subject that even strong-stomached biologists have a hard time studying it. But the killing of infants by newly ascendant males has been documented in lions, leopards, mice, prairie dogs, and a number of nonhuman primates, including chimpanzees.[57] In fact, by now it has been found in so many polygynous species that whenever a new one is observed, and male takeovers are reported, it is assumed that infanticide will also occur. Nearly always it does. Although a nursing infant is hardly a match for a powerful adult male, not uncommonly (and not surprisingly), these males have to work hard to achieve their goal. After all, mothers are selected to resist the murderous males—at least up to a point—and are often supported in defending their offspring by other relatives, mostly females who may themselves be nearing the end of their reproductive careers (and therefore have less to lose by risking injury or death) and who share genes with the endangered babies.

Most of the time, however, infanticidal males are successful, which adds yet another gruesome chapter to the tale of how and why polygynous males can be evolutionarily rewarded for interpersonal violence. Moreover, there is abundant evidence that human beings are not immune to the siren, sociopathic call of infanticide. Once again, the husband and wife team of Martin Daly and Margo Wilson have led the way, showing that severe child abuse, maltreatment, and yes, infanticide are all strongly correlated with the presence of a nonbiological parent. Their findings—to condense an immense amount of data into a simple statement—are that far and away the greatest risk factor identified for children was living with a nonbiological "parent."[58]

This is not to say that step parents are necessarily abusive or infanticidal, but growing up in a family with a stepparent goes farther toward predicting such an outcome than do such considerations as household income, race, religious orientation, geographic location, and so forth. As Daly and Wilson conclude, "Step-relationship remains the single most important risk factor for severe child maltreatment yet discovered."[59]

Among the numerous blood-curdling proclamations found in the Old Testament, demands to kill the infants ring especially savage in the modern ear, although they are sadly congruent with evolution's handwork. For example, "Now therefore kill every male among the little ones, and kill every woman who has known man lying with him. But all the young girls who have not known man lying with him, keep alive for yourselves" (New American Standard Bible, Num. 31:17–18).

Infanticide could happen in the context of serial monogamy as well but is probably more pronounced given polygyny, where the stage is set not only for male-initiated infanticide but for women to give vent to the near-universal problem of the "evil stepmother." Nor is this limited to serial monogamy, as in the tales of Cinderella, or Snow White. After all, in a very real sense, co-wives are stepparents to each other's children, which may well contribute to the tribulations of women—and especially their children—in ongoing polygynous situations (chapter 4).

What about female–female violence? We know remarkably little about it, except that it is less frequent and less overtly violent than its male–male counterpart. Sarah Hrdy argues that much of woman–woman competition is subtle and as a result, often goes unnoticed:

Consider the problem that a human ethologist would face [measuring] such phenomena as sisters-in-law vying for a family inheritance which is to be passed on to their respective children, or the competition for status between mothers who perceive, however dimly, that their own "place in society" . . . may determine the rank at which their own children enter the community at maturity. The quantitative study of such behavior in a natural setting hardly exists. We are not yet equipped to measure the elaborations upon old themes that our fabulously inventive, and devious, species creates daily. . . . How do you attach a number to calumny? How do you measure a sweetly worded put-down?[60]

Compared to male–male violence, which is often physical, woman–woman "violence" is more liable to be verbal, intended to demean and derogate the opponent, and not uncommonly aimed at social exclusion rather than physical injury. This is to be expected, given that women are more likely to be involved in social networks, not only with their offspring, but, in cases of polygyny, with co-wives and their offspring.[61]

Cross-cuturally, 15 year-old girls are liable to use "indirect aggression" more than twice as often as are same-age boys. A recent research study found, in summary, that

> Human females have a particular proclivity for using indirect aggression, which is typically directed at other females, especially attractive and sexually available females, in the context of intrasexual competition for mates. Indirect aggression is an effective intrasexual competition strategy. It is associated with a diminished willingness to compete on the part of victims and with greater dating and sexual behaviour among those who perpetrate the aggression.[62]

It is also worth noting that when derogating other women to other women, such indirect aggression typically involves criticizing the victim's appearance as well as seeking to generate social exclusion, whereas when speaking *to men*, same-sex competitors are especially likely to allege that their rival is a slut.[63] This tactic is especially effective when the men in question have been seeking a long-term relationship, whereas it has the opposite effect on men seeking short-term sexual contact (surprise?).[64]

In the next chapter, we turn to some realities of human sexuality per se, where, as in the case of violence, we will find that the ghosts of polygamy are very much alive.

Notes

1. James, W. (1891). *The principles of psychology*. London: Macmillan.
2. Courtiol, A., Raymond, M., Godelle, B., & Ferdy, J. B. (2010). Mate choice and human stature: Homogamy as a unified framework for understanding mating preferences. *Evolution, 64*, 2189–2203.
3. Toma, C. L., Hancock, J. T., & Ellison, N. B. (2008). Separating fact from fiction: An examination of deceptive self-presentation in online dating profiles. *Personality and Social Psychology Bulletin, 34*, 1023–1036.
4. Buss, D. M., Abbott, M., Angleitner, A., Asherian, A., Biaggio, A., Blanco-Villasenor, A., . . . Yang, K.-S. (1990). International preferences in selecting mates: A study of 37 cultures. *Journal of Cross-Cultural Psychology, 21*, 5–47.
5. Wilson, M., & Daly, M. (1985). Competitiveness, risk taking, and violence: The young male syndrome. *Ethology and Sociobiology, 6*, 59–73.
6. Ryan, C., & Jetha, C. (2011). *Sex at dawn*. New York: Harper.
7. Chimpanzee and bonobo social groups may or may not constitute a "horde," but there is no reason whatever to think that they are, or were, "primal." Human beings did not

evolve from either chimps or bonobos, and we know literally nothing about the mating system(s) of our shared ancestors except that the overwhelming likelihood is that as mammals, they showed the typical mammalian predisposition to overt polygyny, compounded with covert polyandry.

8. de Waal, F. B. M. (2001). Apes from Venus: bonobos and human social organization. In F. B. M. de Waal (Ed.), *Tree of origin: What primate behavior can tell us about human social evolution.* Cambridge, MA: Harvard University Press.

9. Emlen, D. (2014). *Animal weapons: The evolution of battle.* New York: Henry Holt.

10. Daly, M., Wilson, M., & Weghorst, S. (1982). Male sexual jealousy. *Ethology and Sociobiology, 3*, 11–27.

11. Weinstein J. D. (1986). *Adultery, law, and the state: A history.* Hastings Law Journal. San Francisco: University of California.

12. Bakken, G. M. (Ed.). (2000). *Law in the Western United States.* Norman: University of Oklahoma Press.

13. Ibid.

14. "Husband Allegedly Kills Wife's Lover; Wife Indicted." ABC NEWS. http://abcnews.go.com/GMA/LegalCenter/story?id=2998744.

15. Barash, D. P., & Lipton, J. E. (2002). *Gender gap: The biology of male-female differences.* New Brunswick, NJ: Transaction Publishers.

16. Buss, D. M. (1994). *The evolution of desire: Strategies of human mating.* New York: Basic Books.

17. Daly, M., Wilson, M., & Vasdev, N. (2001). Income inequality and homicide rates in Canada and the United States. *Canadian Journal of Criminology, 43*, 219–236.

18. Chagnon, N. (1979). Is reproductive success equal in egalitarian societies? In N. Chagnon & W. Irons (Eds.), *Evolutionary biology and human social behavior.* North Scituate, MA: Duxbury Press.

19. Betzig, L. (2008). *Despotism and differential reproduction.* New Brunswick, NJ: Aldine Transaction.

20. Betzig, L. (2012). Means, variances, and ranges in reproductive success: Comparative evidence. *Evolution and Human Behavior, 33*, 309–317.

21. Chagnon, N. (1988). Life histories, blood revenge and warfare in a tribal population. *Science, 239*, 985–992.

22. McElvaine, R. S. (2001). *Eve's seed: Biology, the sexes and the course of history.* New York: McGraw-Hill.

23. Interesting, isn't it, that the word "conquest" is used not only for wartime success but sexual, too. My guess—unverified at present—is that this usage is pretty much restricted to men.

24. Betzig, *Despotism and differential reproduction.*

25. Zerjal, T., Xue, Y., Bertorelle, G., Wells, R. S., Bao, W., Zhu, S., . . . Tyler-Smith, C. (2003). The genetic legacy of the Mongols. *The American Journal of Human Genetics, 72*, 717–721.

26. Rodzinski, W. (1979). *A. History of China.* Oxford, England: Pergamon.

27. Wade, N. (2006). *Before the dawn: Recovering the lost history of our ancestors.* New York: Penguin.

28. Jarvis, P. (2006). "Rough and tumble" play: Lessons in life. *Evolutionary Psychology, 4*, 330–346.

29. Maccoby, E. E., & Jacklin, C. N. (1974). *The psychology of sex differences.* Stanford, CA: Stanford University Press; Maccoby, E. E., & Jacklin, C. N. (1980). Sex differences in aggression: A rejoinder and reprise. *Child Development, 51,* 964–980.

30. Huck, M. G., & Fernandez-Duque, E. (2011). Building babies when dads help: Infant development of owl monkeys and other primates with allo-maternal care. In K. Clancy, K. Hinde, & J. Rutherford (Eds.), *Building babies: Proximate and ultimate perspectives of primate developmental trajectories.* New York: Springer Verlag.

31. Alexander, G. M., & Hines, M. (2002). Sex differences in response to children's toys in nonhuman primates (*Cercopithecus aethiops sabaeus*). *Evolution and Human Behavior, 23,* 467–479.

32. Chau, M. J., Stone, A. I., Mendoza, S. P., & Bales, K. L. (2008). Is play behavior sexually dimorphic in monogamous species? *Ethology, 114,* 989–998.

33. Geary, D. C. (2010). *Male, female: The evolution of human sex differences.* Washington, DC: American Psychological Association.

34. Patterson, G. R., Littman, R. A., & Bricker, W. (1967). Assertive behavior in children: A step toward a theory of aggression. *Monographs of the Society for Research in Child Development, 32,* 1–43.

35. Bjerke, T. (1992). Sex differences in aggression. In J. M. G. van der Dennen (Ed.), *The nature of the sexes.* Groningen, Holland: Origen Press.

36. Freedman, D. G. (1975). *Human infancy.* New York: John Wiley & Co.

37. Konner, M. (2010). *The evolution of childhood.* Cambridge, MA: Harvard University Press.

38. Archer, J. (2004). Sex differences in aggression in real-world settings: A meta-analytic review. *Review of General Psychology, 8,* 291–322.

39. Whiting, B. B., & Whiting, J. W. (1975). *Children of six cultures: A psycho-cultural analysis.* Cambridge, MA: Harvard University Press.

40. Konner, M. (2015). *Women after all.* New York: W. W. Norton.

41. Sapolsky, R. M. (1998). *The trouble with testosterone.* New York: Scribner's.

42. Mead, M. (1949). *Male and female.* New York: Morrow.

43. Williams, J. E., & Best, D. (1982). *Measuring sex stereotypes: A thirty nation study.* London: Sage.

44. Pfaff, D. W. (2010). *Man and woman: An inside story.* New York: Oxford University Press.

45. Clutton-Brock, T. H., & Isvaran, K. (2007). Sex differences in ageing in natural populations of vertebrates. *Proceedings of the Royal Society of London, B 274,* 3097–3104.

46. Kruger, D. J., & Nesse, R. M. (2006). An evolutionary life-history framework for understanding sex differences in human mortality rates. *Human Nature, 17,* 74–97.

47. Wilson, C. G. (2008). Male genital mutilation: An adaptation to sexual conflict. *Evolution and Human Behavior, 29,* 149–164.

48. Davis, A. J. (1970). Sexual assaults in the Philadelphia prison system. In J. H. Gagnon and W. Simon (Eds.), *The sexual scene.* Chicago: Aldine.

49. Manson, J., & Wrangham, R. W. (1991). Intergroup aggression in chimpanzees and humans. *Current Anthropology, 32,* 369–390.

50. Bruce, N., & Gladstone, R. (2014, December 11). More people die from homicide than in wars, U.N. says. *New York Times,* p. A11.

51. Daly, M., & Wilson, M. (1988). *Homicide.* Chicago: Aldine Transaction.

52. Wolfgang, M. (1958). *Patterns in criminal homicide.* Philadelphia: University of Pennsylvania Press.

53. Matthiessen, P. (2012). *Under the mountain wall.* New York: Random House.

54. Lewis, O. (1961). *The children of Sanchez: Autobiography of a Mexican family.* New York: Random House.

55. Hrdy, S. B. (1980). *The langurs of Abu: Female and male strategies of reproduction.* Cambridge, Mass.: Harvard University Press.

56. Hrdy, S. B. (1977). Infanticide as a primate reproductive strategy. *American Scientist 65,* 40–49.

57. van Schaik, C. P., & Janson, C. H. (2000). *Infanticide by males and its implications.* Cambridge, England: Cambridge University Press.

58. Daly, M., & Wilson, M. (1985). Child abuse and other risks of not living with both parents. *Ethology and Sociobiology, 6,* 155–176.

59. Daly, M., & Wilson, M. (1996). Violence against step-children. *Current Directions in Psychological Science, 5,* 77–81.

60. Hrdy, S. B. (1981). *The woman that never evolved.* Cambridge, MA: Harvard University Press.

61. Benenson, J. F., Hodgson, L., Heath, S., & Welch, P. J. (2008). Human sexual differences in the use of social ostracism as a competitive tactic. *International Journal of Primatology, 29,* 1019–1035.

62. Vaillancourt, T. (2013). Do human females use indirect aggression as an intrasexual competition strategy? *Philosophical Transactions of the Royal Society B: Biological Sciences, 368,* 20130080.

63. Buss, D. M., & Schmitt, D. P. (1993). Sexual strategies theory: An evolutionary perspective on human mating. *Psychological review, 100,* 204.

64. Schmitt, D. P., & Buss, D. M. (1996). Strategic self-promotion and competitor derogation: Sex and context effects on the perceived effectiveness of mate attraction tactics. *Journal of Personality and Social Psychology, 70,* 1185.

3

Sex

They still tell the story in parts of rural New Zealand. In the late 19th century, an Episcopalian missionary was being entertained at a Maori village. After feasting and speech-making, and as the participants were about to retire for the night, the local headman—wanting to show maximum hospitality to his distinguished guest—shouted "A woman for the bishop." Seeing the look of disapproval on the prelate's face, he shouted again, even louder: "*Two women* for the bishop!"

He was an astute judge of male sexuality.

Males are biologically defined as sperm makers, females as egg makers. As we saw in chapter 1, polygyny is mostly a consequence of this distinction: in short, one male can fertilize many females, whereas each female can only be fertilized by one male at a time. Additional matings (by the same male or different ones) are unlikely to result in more children. Just as this difference, and the male–male competition it produces, drives the maleness of violence, it has also generated a male penchant for multiple sexual partners, to a degree greatly exceeding the predilection of most females. "Among all peoples, everywhere in the world," concludes the Kinsey Report, "it is understood that the male is more likely than the female to desire sexual relations with a variety of partners."[1]

The ultimate evolutionary basis for this difference is as clear as the beard on Charles Darwin's face. Male fitness—not physical fitness but its evolutionary aspect of success in projecting genes into future generations—can be increased by copulation with a new partner, each of whom, in theory, can be inseminated as a result. By contrast, because females produce relatively few eggs whereas males produce immense numbers of sperm, the likelihood is that "extra-pair copulations" will not comparably increase the reproductive success of sexually adventuring females. Nonetheless, DNA fingerprinting has recently revealed that among many species, including *Homo sapiens*, females are more inclined to sexual variety than had previously been thought: that is, to polyandry.[2]

Although the evolutionary payoff to polygynous males has long been recognized, only recently have biologists begun to identify the various (and much more subtle) upsides to polyandry in women. Men and women are thus more similar than biologists had—until quite recently—expected. On the other hand, and despite the fact that polygyny and polyandry are complementary sides to the same human sexual coin, there are some intriguing differences when it comes to the details of sexual inclinations. In this chapter, I will describe those traits typically sought by each sex when choosing a partner, waiting until later (especially chapter 6) to explore why—given the obvious downsides of getting caught—polyandry nonetheless persists among women.

Seen through the lens of science, sex is even more mysterious than romantics would have you believe. First, we don't even know why it exists at all! Surprisingly, perhaps, the most common answer—reproduction—isn't persuasive. This is because many living things reproduce *asexually*, and by foregoing sex, they avoid a range of liabilities such as the time and energy expended in courtship; the risk that flamboyant sexual attractants, structures, and antics make distracted lovers more vulnerable to predators; the hassle of finding a suitable partner (who must be of the right species, opposite sex, and similarly inclined); the danger of sexually transmitted diseases; as well as the prospect that mating itself—which, in species employing internal fertilization, requires the intimate penetration of one individual by another—could result in physical injury, especially if one partner is bigger, stronger, and more ardent than another.

All these liabilities haven't even broached what is almost certainly the major biological disadvantage of reproducing sexually: the fact that a mother or father can only project one-half of his or her genes into each child, whereas an asexually reproducing parent gets to generate identical copies of his or her entire genotype. Biologists therefore speak of a 50% "cost of meiosis" because an egg or sperm carries only 50% of all possible parental genes; rephrasing this in more accurate, gene-specific language, each gene within an asexual parent is guaranteed to appear in every offspring, whereas "doing it" sexually mandates that each gene has only a 50% chance of making it, evolutionarily.

And yet sex is very much with us, along with most other "higher organisms." Although no one knows why, at least not for certain, current consensus is that sexual reproduction makes up for its numerous liabilities by the flip side of that "cost of meiosis" drawback: producing offspring who are genetically somewhat different from either parent. This greater genetic diversity has evidently been positively selected because environments have always changed, and parents who simply pumped out identical carbon copies[3] of themselves were selected against. Breeding asexually would have been equivalent to buying a hundred lottery tickets, all with the same number: your options for winning are substantially reduced.

Sex also presents another mystery. Why are there two sexes? Again, no one knows. If the adaptive value of sexual reproduction is the generation of genetic diversity, which is accomplished by shuffling and mixing one's genes with those of another, then it is only necessary for two individuals to meet and exchange genes. There needn't be any discernible sexes at all. But essentially all sexually reproducing organisms divide themselves at least into "mating types," with individuals of one type capable of exchanging genes—that is, "having sex"—with any other individual provided that the two are from different strains. *Tetrahymena thermophila*, a protozoan, has seven mating types. *Schizophyllum commune*, a mushroom, has over 28,000.[4]

If we interpret sex as who-can-mate-with-whom for the purpose of recombining genes, then there can in theory be a large number of sexes indeed, and such a world would seem to offer a dazzling array of opportunities. But for some unknown reason, most of the sexually reproducing, organic world has made the evolutionary "decision" to specialize as either egg maker or sperm maker, which narrowly structures the reproductive landscape within which we and all other mammals must make the best of things.

But I don't mean to sound glum. Evolution also has the delightful habit of using pleasure as a lure, getting its products to enjoy those things that are good for them: eating when hungry, sleeping when tired, deriving pleasure from productive activities . . . including sex. At the same time, unfortunately, our stubborn biological insistence on segregating ourselves into two distinctly different sexes—each with somewhat different agendas—has set the stage for some real trouble in that erotic paradise. Both men and women have evolved wanting, in a sense, the same thing—reproductive success, or "fitness"—but given their differing specializations as egg makers and sperm makers (in addition to the fact that individuals are composed of differing genotypes), males and females aren't necessarily on the same page, and reproduction isn't always a positive sum game.

Let's start with the Rodgers and Hammerstein musical *The King and I.* Based on *Anna and the King of Siam*, a novel by missionary Margaret Landon, it was in turn derived from a memoir, *The English Governess at the Siamese Court*, by Anna Leonowens, an Englishwoman who served as governess to the children of King Mongkut of Siam in the early 1860s. In the musical, the King and Anna, although mutually in love, have a sometimes tragic, sometimes comic disagreement when it comes to the merits of monogamy versus polygamy, and accordingly, between a stereotype of male versus female sexual inclinations:

THE KING: A woman is designed for pleasing man, that is all. A man is designed to be pleased by many women.

ANNA: Then how do you explain, your majesty, that many men remain faithful to one wife?

THE KING: They are sick.

ANNA: Oh, but you do expect women to be faithful?

THE KING: Naturally.

ANNA: Well, why naturally?

THE KING: Because it is natural. It is like old Siamese saying, "A girl is like a blossom, with honey for just one man. A man is like a honey bee and gather all he can. To fly from blossom to blossom a honeybee must be free. But blossom must never fly from bee to bee to bee."

ANNA: Ha ha. Oh your majesty in England we have a far different attitude. We believe for a man to be truly happy he must love one woman and one woman only.

THE KING: This idea was invented by woman.

This idea—monogamy—almost certainly wasn't invented by natural selection. (At least not for human beings.) But as we'll see, neither was the "idea" that women, too, require monogamy, although the evidence is pretty strong that women have been encouraged, both by society and biology, to proclaim their sexual fidelity, while behaving polyandrously, but on the sly. As to why women would be especially inclined to under-emphasize their sexual opportunism, just consider the reality of violent male sexual jealousy described in chapter 2.

Men's polygyny conflicts vigorously—although often subtly—with women's polyandry. A naïve view would predict that men's and women's sexual proclivities don't necessarily conflict at all, and both could readily be incorporated within one happy, swinging, multimale and multifemale social group, a blissful polyamorous unit in which everyone who wanted to had sex with everyone else who wanted to. But as we've seen, polygynandry simply doesn't work for human beings (or nearly any other species either).

Natural selection induces each sex to seek—or at least, to be susceptible to—multiple partners, while at the same time motivating each sex to prefer, often violently so, that his or her identified partner not behave similarly. The result is trouble, whether the official mating system be polygyny, polyandry, or monogamy. Hence, as Shakespeare noted in *A Midsummer Night's Dream*, "The course of true love never did run smooth." In fact, those who claim that sexual jealousy doesn't exist or that it can easily be overcome, that polyamory is "natural" and free of difficult conflicts, are either very far out on the normal curve that describes nearly all human beings or downright lying.

L et's start once again with polygyny. The male preference for multiple sexual partners is evident, as we've already noted, from the great majority of different human social groups in which polygyny is permitted for those men able to attain it. When a serf was married in medieval Europe, for example, the lord of the manor had the right—the *droit du seigneur*—to spend the wedding night with the new bride. Historians have never seen fit to ask why it was the lord who had sex with the blushing bride and not the lady of the manor who initiated the bashful groom. There never was a *droit de la madame*. Although some royal ladies undoubtedly managed secretive extramarital liaisons when they had the will and the opportunity—much as women do today—such behavior was likely not a published right as much as a risky endeavor (as Camelot's Queen Guinevere found out).

In his now-classic research, geneticist A. J. Bateman concluded that "an almost universal attribute of sexual reproduction" is the "greater dependence of males for their fertility on frequency of insemination," which in turn results in an "undiscriminatory eagerness in the males and a discriminatory passivity in the females."[5] As a result, males typically court females and not the other way around. Bateman also concluded that because additional matings benefit males to a much greater degree than they do females, there is a "greater male inclination to polygamy." Moreover, even among species that have evolved monogamy—at least social monogamy—this difference is one that endures.

To be sure, women, too, want sex, and doubtless always have. But they are far less pushy about it, often providing sexual favors in return for gifts or as a concomitant of intimacy. The pattern is so well established that the phrase "sexual favors" immediately suggests the dispensing of erotic goodies by women, to men, not vice versa. In the first century AD, the Roman philosopher Lucretius wrote a book with the marvelously modest title, *On the Nature of the Universe*, in which he presented his vision of the prehistoric human sex drive:

> They lived for many revolutions of the sun, roaming far and wide in the manner of wild beasts. And Venus joined the bodies of lovers in the forest; for they were brought together by mutual desire, or by the frenzied force and violent lust of the man, or by a bribe of acorns, pears, or arbute-berries.

There can be little doubt that it is the men who bribed the women, offering food in return for sexual access.

It is also striking how often, in the modern Western world, when a new religious cult is formed, the typically charismatic male founder proclaims his "right" or even "obligation" to mate with a large number of the group's women. At least as striking is the willingness with which these women—and even more so, the excluded men—acquiesce: testimony, perhaps, to our long history of adjusting to a regime of harem-holding dominant males. In chapter 7, we encounter the possibility that this long history may be intimately connected to the spread of monotheism.

Although the sexual motivation of polygynous alpha males is nearly always downplayed in accounts of the daily practice of cultists, it stretches credulity to imagine that the men in charge weren't responding to their own polygynous inclinations. The Church of Jesus Christ of Latter Day Saints (i.e., Mormons) have been more than a little embarrassed by their polygynous history and have only recently owned up to it. But it has long been known that Joseph Smith had something like 30 wives (by some accounts, as many as 60), and his disciple Brigham Young had 51. Not surprisingly, the wives to whom these leaders were "spiritually sealed"—and with whom they were sexually active—were nearly always young and thus not only physically attractive but reproductively competent at the time of their marriages.

David Koresh, who made himself leader of the Branch Davidian cult in Waco, Texas, also conferred on himself sexual access to all the women—including young teenage girls—among his followers. During the brief time of that group's existence, he is believed to have fathered 15 children.[6] Jim Jones, who started the ultimately suicidal People's Temple, prohibited extramarital sex among his followers . . . except for himself. There may be examples of cult leaders who did *not* make use of their position to further their personal polygyny, but I cannot think of any.

Polygyny wasn't only the perquisite of human cult leaders. The Greek gods were a notably randy lot. Zeus, for example, fathered many of the lesser gods by coitus with Titans; indeed, he had sex with just about anything that moved, including many human females (always to Hera's displeasure, although she was notably unable to do much about it). I can't identify a male Greek god who *didn't* have multiple sexual partners. The same was true, if anything more so, within the Hindu pantheon. In chapter 7, we briefly explore the extent to which the Judeo-Christian-Islamic God, who isn't above identifying himself as "jealous," also fits the mold of a polygynous alpha male.

According to author Hector Garcia

> The Abrahamic God has also been known to womanize. In Ezekiel 23 Yahweh is described as having two women—Samaria and Jerusalem. Here He personalizes these entire cities as women in his life, a subject to which we will return. Notably, however, there were not one but two wives. In addition, there are scores of patriarchs in the Old Testament who were polygamous, including: Abraham, Abijah, Ahab, Ashur, Belshazzar, David, Elkanah, Esau, Gideon, Hosea, Jacob, Jehoram, Joash, Jehiachin, and Rehaboam. King Solomon trumps them all . . . with 700 wives and 300 concubines (1 Kings 11:1–3).[7]

Recall the mantra from early in the Watergate investigation: "Follow the money." When it comes to religious cults, and perhaps even mainstream practice, "Follow the sexual access." When you do, you are likely to find that the social systems typically established by men provide that these men end up having sex with numerous women. Those religions that officially practiced celibacy among their leadership—such as the Roman Catholic Church and some sectors of Buddhism—may or may not have actually limited their reproduction as advertised.

The harem-acquiring proclivity of men isn't limited to newly established religions or cults. Natural selection has endowed male sexuality with a particular lure that reflects the reality of being a sperm maker while also stimulating those sperm makers to have sex with multiple partners when possible. It is known as the Coolidge effect because of the following story (which may well be apocryphal, although the phenomenon it describes definitely is not).

It seems that President and Mrs. Coolidge were on separate tours of a model farm when Mrs. Coolidge noticed that the one rooster was mating quite frequently. She asked if this happened often and was told "many times, every day," whereupon she asked that the president be told this when he came by. Duly informed, President Coolidge asked, "Same hen every time?" The reply was, "Oh, no, Mr. President, a different hen every time." The president answered: "Tell *that* to Mrs. Coolidge."

It turns out that roosters also ejaculate greater quantities of sperm when paired with a new hen as compared with a sexually familiar one.[8] The technical report of these findings was titled "Sophisticated Sperm Allocation in Male Fowl"; and there is some evidence that humans, too, engage in comparably "prudent" sperm allocation, ejaculating more sperm when with a new partner and more during sexual intercourse than during masturbation.[9]

The term Coolidge effect was coined by Frank Beach, a renowned comparative psychologist and early researcher on sexual behavior of animals and people. It applies to a large number of birds and mammals: not just barnyard poultry, but horses, cows, pigs, goats, and sheep. Introduce a ram to a ewe in heat and he will mount her. Then perhaps he will again. After a short while, however, his sexual appetite will peter out. But a new ewe makes for a new him.

The same thing happens to human beings, and is for men both an indicator of polygyny, and a stimulant thereof. There is essentially no female equivalent, although fantasies persist, such as the reputed appeal of the "tall, dark stranger." A summer romance or equivalent can definitely be attractive to women no less than to men, but note the emphasis on "romance" rather than what Erika Jong memorably called a "zipless fuck." When it comes to sexual stimuli, a novel partner itself does not seem to exert anything like the erotic appeal to women that it does to men.

It is a fascinating irony that although men stand to gain more—in terms of producing offspring—from multiple copulations, women are physiologically capable of "having" more sex than men, such that one woman could sexually satisfy multiple husbands in a single night and continue to do so repeatedly, whereas one man cannot perform similarly with regard to multiple wives. Yet, as we have seen, social systems are commonly structured the other way around, with polyandry being exceedingly rare. In his *Letters From the Earth*, Mark Twain had great fun with this paradox. Here is Twain's Devil reporting his discoveries, after visiting our planet:

> Now there you have a sample of man's "reasoning powers," as he calls them. He observes certain facts. For instance, that in all his life he never sees the day that he can satisfy one woman; also, that no woman ever sees the day that she can't overwork, and defeat, and put out of commission any ten masculine plants that can be put to bed to her. He puts those strikingly suggestive and luminous facts together, and from them draws this astonishing conclusion: The Creator intended the woman to be restricted to one man.
>
> Now if you or any other really intelligent person were arranging the fairnesses, and justices between man and woman, you would give the man a one-fiftieth interest in one woman, and the woman a harem. Now wouldn't you? Necessarily, I give you my word, this creature with the decrepit candle has arranged it exactly the other way.

Although Twain's Devil is absolutely right when it comes to sexual physiology, nonetheless, from an evolutionary perspective, it is more logical for one man to mate with multiple women than to have one woman mated to several men. And in this case, evolutionary bio-logic has won out—at least when it comes to the majority of socially approved reproductive arrangements.

Among the many other instances of such bio-logic, many studies have looked at male–female differences in behavior, with perhaps the most consistent result involving distinct differences when it comes to desire for sexual variety. Although it is clear that women, too, can be interested in a diversity of sexual partners—and are typically inhibited via cultural tradition from admitting it—among human beings worldwide, men are significantly more interested. Here is a summary of this pattern, slightly modified, from a recent volume titled *Mating Intelligence Unleashed: The Role of the*

Mind in Sex, Dating and Love[10] (in this excellent book, each of the following statements is supported by references to the technical literature):

1. Men across the globe in 48 nations report an interest in having more lifetime sex partners compared with women.
2. Men in 53 nations show higher levels of sociosexuality (a proxy for promiscuity) than women, and women were more variable than men in their sex drive.
3. Men select more partners in a speed-dating context.
4. When asked how early in a relationship it would be okay to have sex, men's answers came in much earlier compared with women's answers.
5. Men report more reasons for having sex, and their reasons centered more on physical appearance and physical desirability.
6. Men are more likely than women to engage in extradyadic sex (i.e., sex outside of a relationship).
7. Men are more likely than women to be sexually unfaithful multiple times with different sexual partners.
8. Men are more likely than women to seek short-term sex partners who are already married.
9. Men are more likely than women to have sexual fantasies involving short-term sex and multiple opposite-sex partners.
10. Men are more likely than women to pay for short-term sex with (male or female) prostitutes.
11. Men are more likely than women to enjoy sexual magazines and videos containing themes of short-term sex and sex with multiple partners.
12. Men are more likely than women to desire, have, and reproductively benefit from multiple mates and spouses.
13. Men desire larger numbers of sex partners than women do over brief periods of time.
14. Men are more likely than women to seek one-night stands.
15. Men are quicker than women to consent to having sex after a brief period of time.
16. Men are more likely than women to consent to sex with a stranger.
17. Men have more positive attitudes than women toward casual sex and short-term mating.
18. Men are less likely than women to regret short-term sex or "hook-ups."
19. Men have more unrestricted sociosexual attitudes and behaviors than women.
20. Men generally relax mate preferences (whereas women increase selectivity for physical attractiveness) in short-term mating contexts.
21. Men interpret that strangers have more sexual interest in them than women do.

These generalizations having been stated, I need to emphasize that although they are statistically valid, they are just that: generalizations, which is to say, statements that are *generally* true. Individuals differ greatly, the specifics influenced not only by personal differences but also by local cultural rules and circumstances: especially when it comes to expectations about what can be described as sexual freedom, permissiveness, or—more neutrally—attitudes. In this regard, Norway may be particularly revelatory because compared to most, Norwegian society has a great deal of gender equality. Norwegian women report that they would like to have approximately two different sex partners during the coming year; Norwegian men, seven.[11]

"Why can't a woman," exclaims Henry Higgins, in *My Fair Lady*, "be more like a man?" In many ways, she can (and vice versa). But when it comes to some of our most intimate, unconscious, and biologically relevant details, women and men really are different. And these differences are far more intriguing than mere plumbing—although, as we have seen, aspects of that plumbing drive the more subtle behavioral distinctions. A consistent male–female difference, in addition to preference for multiple sex partners, is that whereas men are especially prone to evaluate women by their youth and physical appearance, women are especially prone to evaluate men by what biologists call "resource control," and what normal people mean by wealth, status, power, and so forth.

Among the Hadza of Tanzania, for whom large game comprise a much valued resource, women are especially impressed with a man's prowess as a hunter.[12] This pattern continues to hold in the Amazonian rainforest no less than in East Africa: in a community surrounding Conambo, Ecuador, consisting of a small-scale hunting and gathering society, women's assessment of male attractiveness was particularly influenced by "warriorship" and hunting skills.[13]

In a now-classic study, evolutionary psychologist David Buss found that in a cross-cultural sample examining 37 different human societies, women consistently chose mates based on their financial wherewithal, ambition, and industriousness, whereas men consistently looked for physical attractiveness and relative youth.[14] "Women more than men in all 37 cultures valued potential mates with good financial prospects," he reported. "Men more than women across the globe placed a premium on youth and physical attractiveness, two hypothesized correlates of fertility and reproductive value."

A raft of studies have subsequently examined male–female differences when it comes to online dating advertisements, and the findings have been consistent: men announce their wealth and income more than do women, who more often point to their youth and attractiveness. And not surprisingly, women correspondingly react more positively to indications of a man's resources.[15]

Readers of a certain age may recall the following folk song, which enjoyed renewed popularity in the 1960s: "If I were a carpenter. And you were a lady, Would you marry me anyway? Would you have my baby?" Romantic fantasy aside, the great majority of ladies, in the great majority of cultures, would unhesitatingly answer "No way!"

The bottom line, confirmed by Buss and his colleagues numerous times, is that men are more focused on physical attributes; women, on social indicators. In one study, women college students were presented with men, some of whom were wearing Rolex watches and spiffy clothing and others a Burger King uniform. When they were asked which men would they prefer to talk with, date, have uncommitted sex with, or consider marrying, the results were so consistent and predictable that I won't even repeat them here.[16] So why mention them at all? Because when some things are so obvious as to evoke a "Duh!" response, this says something about them being so true that they are in danger of being overlooked.

One possible explanation for these differences is that women place an inordinately high value on male control of resources (car, career, etc.) as a result of women's poverty and dependent circumstance, rather than some biologically mediated predisposition. However, a study of personal ads in Sweden—a country in which, like most Scandinavian societies, the economic and social situation of women is essentially equal to that of men—women nonetheless specify the importance of resources three times more than do men.[17]

Interestingly, women aren't simply focused on male wealth; *how* a man came by his resources also matters, with women significantly more interested in men who have earned their money rather than inherited it.[18] This presumably speaks to female concern about male personality traits (ambition, intelligence, work ethic) as opposed to a simple "gold-digger" stereotype. Also worth noting is that this concern is especially apparent in the context of potential long-term mating rather than a one-night stand.

Social scientists have long had a very hard time acknowledging the role of biology when it comes to pretty much anything (except perhaps who sits down and who stands up when urinating). For example, the pioneering social anthropologist Bronislaw Malinowski, describing sexual relations among the Trobriand Islanders of the South Pacific, was hard-pressed to admit that there was a distinct asymmetry between men and women when it came to sexual intercourse and that rather than being reciprocal, it was "a service from women to men" for which "men have to pay." In discussing this, Malinowski felt the need to assert that this male–female asymmetry results from the impact of what he called "custom, arbitrary and inconsequent here as elsewhere," which in his opinion has arbitrarily "decreed" this male–female distinction.[19]

The pattern of sex differences that Malinowski describes and then seeks to attribute solely to the culturally generated particulars of a specific group of people is conspicuously and consistently *cross*-cultural. This would hardly be expected if the tendencies in question are a result of arbitrary and inconsequential custom, in which case there should be as many societies where, for example, women pay men for sex as vice versa, or it should be equally common for men to assess a woman's sex appeal as a function of her wealth and status as it is the other way around.

A cross-cultural review concluded that—even in industrial societies—not only is sex something that men purchase and women sell (or are coerced into providing), but that the higher a man's status and the greater his wealth the more likely he is to have obtained more than one wife; that is, to be polygynously mated.[20] And at the other end of the socioeconomic scale, the situation is inverted. Early in his career, George Orwell spent time living the "low life" and then writing about it in his memoire, *Down and Out in Paris and London*. Orwell sounds like an evolutionary psychologist when he observes that "any presentable woman can, in the last resort, attach herself to some man," after which he describes the situation of the "tramp," who is condemned to perpetual celibacy. For of course it goes without saying that if a tramp finds no women at his own level, those above—even a very little above—are as far out of his reach as the moon. The reasons are not worth discussing, but there is no doubt that women never, or hardly ever, condescend to men who are much poorer than themselves. A tramp, therefore, is a celibate from the moment when he takes to the road. He is absolutely without hope of getting a wife, a mistress, or any kind of woman except—very rarely, when he can raise a few shillings—a prostitute.[21]

Echoes of the evolutionary, polygamous theme linger on, even in modern technological societies. In one study of French Canadian men, no correlation was found between socioeconomic status and reproductive success. Yet, when that same study probed a bit deeper, it found that in a precontraceptive society (which was true for most of humanity's evolutionary history), wealthier, more successful men would have had many more children. This was determined by considering the number of copulations as well as the number of sexual partners and estimating the "number of potential conceptions" if birth control had not been used.[22]

A ll of the preceding is consistent with a male predilection for polygyny, along with a female preference for males likely to be successful providers and harem holders. What about female predilection for polyandry? It would seem that the male–female differences described here argue against it. But wait! First, we need to keep in mind that males are generally up front about polygyny because even within societies that don't officially embrace harem establishment, men are culturally encouraged to acknowledge interest in sexual variety (or at least, they are less likely to be criticized for doing so), whereas women are taught to be more sexually discreet. And second, because of these cultural restraints, we need to look more deeply—into unconscious female inclinations and traits—to detect the stirrings of polyandry.

The received wisdom among specialists in reproductive physiology and behavioral endocrinology has been that women—unlike many other female mammals—do not experience estrus, or "heat." The word estrus derives from the Greek *oistros*, which originally referred to "gadfly, sting, and mad impulse." The idea is that animals in estrus behave madly, like the victim of a stinging fly. Clearly, women do not experience regular sexual frenzy. Compared to all other mammals, female human beings exercise an immense amount of control over who they mate with—something that almost surely conveys adaptive advantage in itself.

It has long been argued (by me,[23] for example), that human beings were notable—and perhaps unique—in that women were sexually receptive throughout their ovulatory cycle. This is consistent with monogamy, whereby continuous female receptivity could be in the service of pair bonding, an adaptive tactic by which our great-great-grandmothers kept our great-great-grandfathers from sexually straying. Why look elsewhere, the argument went, when you can be sexually satisfied, any time of the month, at home?

However, an avalanche of recent findings has demonstrated that contrary to what had been the received wisdom for many years, women are not equally responsive sexually at all times of their ovulatory cycle. The take-home message is that women behave differently when they are ovulating, and thus, when they are most fertile. And among these differences are varying preferences for male partners at such times. Perhaps the most dramatic finding involved a study of topless pole dancers in Albuquerque, New Mexico, which looked at the tips received by these women as a function of their menstrual cycles. To the surprise of the researchers—and the women, too—it turned out that they received significantly more money when they were maximally fertile.[24] This could have been due to women acting especially alluring during those days, or perhaps to some currently unidentified pheromone of which men were unconsciously aware, or maybe something else, presently unknown.

Whatever the precise proximate mechanism, women wouldn't be selected to be more sexy when ovulating if they experienced a consistently monogamous mating system. It turns out that ovulation has a substantial impact on women's sexual inclinations. Here is a summary of the most notable (once again slightly modified from *Mating Intelligence Unleashed*). When ovulating, women engage in

1. More female-initiated sex with their partners;
2. A relatively strong preference for traditionally "masculine" partners—such as lower voice, heavier beards, and more evident upper body musculature;
3. A greater likelihood of sexual infidelity if they are engaged in a current relationship;
4. A greater likelihood of touching nearby men in casual social situations;
5. A tendency to be attracted to the scent of men whose bodies are relatively symmetrical;
6. A tendency to be attracted to men who display indications of creativity (musical, artistic, verbal, etc.);
7. A greater inclination to take physical risks and to travel farther from home;
8. A greater interest in erotic movies;
9. A tendency to wear clothing that displays more bare skin; and
10. A tendency to dance with more vigorous and sensuous movements.

An intriguing recent finding (expanding on the preceding number 9) is that women—at least in Western countries—are more likely to wear pink or red when they are ovulating (most fertile). However, this "red dress effect" occurs only in winter, not in summer, presumably because in warmer weather, women can draw attention to their bodies by wearing skimpier clothing rather than bright colors.[25]

In turn, these behavioral changes on the part of ovulating women have definite impacts on men, as follows:

1. As already described, men give bigger tips to female strippers who are ovulating;
2. Men find the voices of ovulating women relatively attractive;
3. Men find the scent of ovulating women relatively attractive; and

4. Men report finding photos of ovulating women physically attractive compared with photos of the same women when not ovulating.

The significance of these ovulatory effects can be debated. Clearly, the changes that women experience when ovulating are subtle or else they would have been common knowledge long before evolutionary psychologists discovered them. At least some evolutionary psychologists have nonetheless concluded that women actually do experience estrus.[26] Regardless, it is possible that these changes are fully consistent with a monogamous evolutionary history in which selection would favor women who were predisposed to be especially sexy when they are especially fertile. But there wouldn't be much of a reproductive payoff for women who were monogamously mated to become particularly attractive or responsive when ovulating because they would likely be having intercourse with the same sexual partner in any event. And there certainly wouldn't be selection for an inclination to go away from home, interact more with strange men, and to be more appealing to such men if they weren't also likely to engage sexually with men *other* than their designated partner when they were most likely to conceive (i.e., if they weren't polyandrous).

As a general rule, males of most species spend considerably more time and energy pursuing sexual opportunities than do females, a pattern that almost certainly applies to human beings as well. This assertion is difficult to test quantitatively, however, if only because the techniques used by the two sexes are generally quite different, with men more overt in their search and women more subtle. When a woman insists that she won't go "out" without wearing makeup, is that because she is seeking a sexual opportunity or simply attempting to look her best? And if the latter, does looking "her best" imply anything more than literally putting her best face forward? If so, why bother doing that?

The flip side of estrus—whether purported or real—is concealment of ovulation. And here, human beings are, although not unique, extraordinary among mammals generally and primates in particular. Just a quick visit to a zoo will often reveal a female chimp or baboon sporting an immense rosy red cauliflower on her behind. But don't worry, it's not a tumor; rather, it's her estrus swelling. Women don't have anything like this. Human ovulation is concealed, so much so that even in the medically sophisticated 21st century, it isn't easy to tell when a woman is ovulating. Why is something so biologically important kept so secret?

Hypotheses abound; clear answers, not so much.[27] To begin, it's noteworthy that among other primates, advertised ovulation is associated with a multimale, multifemale lifestyle, perhaps as a way of inducing male–male competition so that a cycling female is fertilized by the best (i.e., most competitively successful) of the many males nearby. As to why human ovulation is concealed, one hypothesis claims that it is part of a "keep him nearby" strategy because a male would be constrained to maintain especially close contact with a female insofar as he didn't know when she was fertile. This would be consistent with the fact that human beings are also notable for the extent to which women remain sexually receptive throughout their cycle, adding to the male's uncertainty and thus restricting his gallivanting after other mates.

Another hypothesis, paradoxically, is almost 180 degrees opposite. Perhaps concealed ovulation is a strategy that permits a secretly cycling female to sneak away and mate with other, peripheral males, precisely when she is most likely to conceive; after all, even a dominant male—especially one whose attention is spread over a bevy of females—cannot guard every female all the time. If a given female clearly indicated her ovulation, she would likely be attended extra closely at that time. By concealing their fertility status, females could be giving themselves greater copulatory opportunities.

Why might they be especially rewarded for mating with peripheral males? Recall the phenomenon of infanticide in polygynous species following a male takeover. It turns out that newly ascendant males are less likely to kill infants born to females with whom they have previously had sexual intercourse. From the male's perspective, this forbearance is a totally adaptive Darwinian strategy, which would make it equally adaptive for females to use extracurricular copulations as a way to take out "infanticide insurance."[28]

Concealed ovulation could be consistent with either monogamy or polygyny, but in any event, it is also consistent with female polyandry insofar as it likely grants females more choice when it comes to mating partners. Even aside from an "estrus-like" tendency to solicit the sexual attention of other men, there is also a distinct evolutionary logic for women who are members of a harem to be particularly enticing when most fertile.

Renowned anthropologist Claude Lévi-Strauss[29] stated that worldwide, marriage is essentially a social contract among men in which women are exchanged as sexual commodities with the goal of solidifying social relationships among men. This highly androcentric, downright patriarchal perspective may well be an exaggeration of the true state of affairs. Nonetheless, it is clear that cross-culturally, men strive vigorously and often conspicuously to control the lives—especially the sex lives—of women. By contrast, although women also have a stake in the sex lives of men, they are less overt about it, and certainly less physically threatening. (Given that men appear to engage in comparatively more philandering than do women, it seems that on balance, women are indeed somewhat less controlling, although once again it is worth emphasizing that when it comes to infidelity by their partners, they have not only substantially less physical and social power, but are also somewhat less at risk of losing fitness.)

In Western cultures, a father traditionally "gives" his daughter in marriage, although historically this is accompanied by a dowry. In most of the world's societies, however, and especially prior to the onslaught of colonialism and imperialism, men purchase their wives via a "bride price." Not only that, but they typically have the right to return their purchase if it proves unacceptable, for example, if the bride isn't a virgin, doesn't conceive children, and so forth.[30] A careful review of more than 200 different societies[31] concluded that almost universally, husbands and their families show particular concern about the fertility of wives, with "barrenness" being a predictable grievance.

Although social scientists have unsurprisingly interpreted such concern as due to its symbolic and social importance, this ignores the literal significance of reproduction—especially for those men making the arrangements. After all, women cannot be duped into thinking that they have reproduced, whereas men can.

It can be argued that large harems are essentially displays of wealth and social power, but this leaves much out, not least the fact that harem members are often chosen as a function of their youth and fertility, that sexual access is often concentrated at times of peak fertility, and so forth. Just as Freud is said to have noted that sometimes a cigar is just a cigar, sometimes a harem is just a harem—a male way of having as many sexual partners, and as many offspring, as possible.

It may seem paradoxical, but evidence for polyandrous proclivities on the part of women can be found by examining the testes and the ejaculate of *men*. Among the highly polygynous gorillas, testes are about .02% of body weight; and gorillas produce a relatively dilute ejaculate, with roughly 5 million sperm per squirt. Those polygynandrous chimpanzees have testes that are proportionally 15 times larger (.3% of body weight), and their ejaculate is 14 times more sperm rich (60 million pollywogs at a time). And human beings? We are appropriately in between, although if anything closer to gorillas: human testes are on average .06% of body weight, and sperm concentration is about 25 million per ejaculate. This pattern is also consistent with data on the average number of male sex partners experienced by females per birth: gorillas, one partner per birth; human beings, 1.1; baboons, 8; bonobos, 9; and common chimps, 13.[32]

Not only do men have relatively small testicles, as moderate polygyny would predict, they also have reduced Sertoli cell mass (these cells help nourish the developing sperm and are sometimes called "nurse" cells). In addition, men possess comparatively small sperm reservoirs, of the sort expected if sperm were to be ejaculated perhaps every few days rather than many times per day, as, for example, in multimale, multifemale troops in which sperm competition is high. Incidentally, there are some species in which males produce substantially fewer sperm per ejaculation. Among pipefish, for example, sperm are numbered by the dozen rather than by the hundreds of million.[33] This makes sense because pipefish females deposit their eggs within the males' brood pouch, where they are fertilized without risk of intrusion by other males and their sperm.

For anyone still seduced by the fantasy of primordial human promiscuity, sperm themselves are something of a giveaway. Spermatozoa consist of three basic parts: the head (containing the man's DNA), a mid-piece containing lots of mitochondria (energy-generating subcellular components), and the tail. That mid-piece has been found to be large in proportion as a species engages in substantial amounts of male–male competition, as among chimpanzees and bonobos. It turns out that not only do we have rather small testes for our body size, but the mid-piece section of human sperm is among the smallest recorded for any primate.[34]

Although our closest living relatives are chimpanzees and bonobos, it is blindingly obvious that we didn't evolve in a context of polygynandrous, multimale, multifemale free-for-alls. Yet more evidence comes from a study of primate immune systems that examined 41 different species, looking at their white blood cell counts. The idea is that with more sexual encounters comes more risk of sexually transmitted diseases, and hence, a need for a more robust immune system. This premise was confirmed by the discovery of a positive relationship between the average number of sexual partners per female and the white cell count.[35] Moreover, the researchers report that "humans align most closely with the gorilla, a polygynous species with low sperm competition, and secondarily with a monogamous gibbon."

Next, we take a look at penises.

If you asked a child, "What's a penis for," he would probably respond, "Peeing." Ask an adult and he or she would likely answer "For sex," or more precisely, "For introducing sperm into the female reproductive tract." True enough, but among many animals, the male's penis serves an additional function: removing semen introduced by a previous male. Some sharks have a double-barreled penis, used to provide a stream of seawater, thereby sluicing out existing sperm before mating occurs. Insects sport a wide array of hooks, reamers, and scooping devices, which do the same thing prior to mating.

It has been suggested by some scholars that the human penis, too, is an instrument of sperm competition whereby it (especially the otherwise mysteriously shaped glans) acts as a kind of "plumber's helper," adapted in conjunction with vigorous thrusting to displace sperm deposited by a prior male.[36] Such an adaptation would seem unnecessary if prehistoric women carefully refrained from mating with males other than their monogamous or polygynous mate. So it is at least possible that the human penis testifies—albeit less clearly than do sperm and testes—to the polyandrous mating inclinations of women.

Regardless of the precise degree of women's polyandry, continuous female sexual receptivity isn't necessary for monogamy to be maintained. Thus, gibbons—which are pretty close to monogamous, although not 100%—have sexual intercourse only after the current offspring is weaned, thus only for a brief time every few years; ditto for many monogamous birds. In fact, there is no consistent trend among animals for monogamous species to be more sexually active (outside of estrus) than are other, polygamous species. So the comparatively continuous sexual receptivity of human females needn't be seen as pointing toward monogamy; it could just as readily be interpreted as an adaptation for competition among members of a harem. Moreover, by itself, regular and frequent sex doesn't necessarily conduce to monogamy: consider those famously lascivious bonobos who copulate (and engage in all sorts of frisky sexual practices, including oral sex plus vigorous homosexual activity) at the drop of a leaf and are definitely not behaving in the service of monogamous unions.

We've already considered some of the more obvious biological factors that distinguish male harem masters from female harem members, and from women as independent sexual agents (i.e., polyandrists, whether overt or—more commonly—covert). Let's take a closer look at how these characteristics interact with culturally mediated traits and mating preferences; in short, time to look at sex appeal, polygamy style.

When it comes to choosing a partner, it isn't clear exactly how polygynous societies compare with monogamous ones, although the likelihood is that there is more mutual assessment in the latter, owing simply to the fact that monogamists are putting all their evolutionary eggs in one basket: that of each other. Moreover, where polygyny prospers, women may have less opportunity to choose, either because they are forced by their families or dragooned by a powerful man into joining his harem. In any event, there are substantial differences between what men and women seek in a long-term mateship.

A study of US college students[37] looked at male–female differences in how importantly the subjects rated "earning capacity" when it came to willingness by these students to engage in four different levels of intimacy and involvement: dating, a single episode of sex, regular sexual relations, and marriage. It was found that women consistently rate earning capacity more highly than do men, for all stages. Interestingly—and to no one's surprise—earning capacity's importance was highest (for both sexes) when it came to marriage.

This shouldn't need saying, but it warrants mention nonetheless, specifically because it is so taken for granted as to be readily overlooked: in all of human history, it has never, ever, been the landless peasant, the struggling laborer, the hard-working down-in-the-dirt farmer or the lowly and anonymous spear carrier in a large army who ends up with multiple wives. When Henry Kissinger noted that "power is the ultimate aphrodisiac," he was no less an astute judge of female sexuality than the Maori chieftain, with whom we began this chapter, was of his male counterparts.

Innumerable studies investigating a huge number of different human societies have nearly always come up with the same fundamental result: dominant and otherwise successful men end up with more lovers, more wives, and more children than do "ordinary" men. To some extent, this is due to the simple fact that—as already noted—those who many nontechnological societies identify as "Big Men" are indeed big men; and as such, they are able to compete physically with other men and to coerce women. But there appear to be other considerations as well, at least some of which involve the actual (albeit perhaps unconscious) preferences of women.

Time for a brief detour to peacocks and peahens, specifically the outlandish tail feathers of the former. These elaborate appurtenances have long been a source of perplexity as well as admiration among biologists, beginning with Darwin but definitely not ending with him. After all, it takes lots of metabolic energy to build all that fancy plumage, which are iconic examples of some of the downsides of sexual reproduction. Darwin suggested that the peacock's highly exaggerated feathers evolved in response to choices made by peahens, although because he didn't have a good idea why they would show such preferences, he settled on some sort of "aesthetic sense" on their part. These days, biologists agree that female choice is definitely involved, but we have a much better idea—actually, two related ideas—why females make such choices. (And as we'll soon see, it is at least possible that women make analogous choices, although—for better or worse—the results when it comes to men's secondary sexual traits are much less dramatic. Some men indeed strut their stuff, but they're not constructed like peacocks.)

As to why female choice occurs, here is idea number one: females choose to mate preferentially with ornamented males because as a result, their own male offspring will be similarly ornamented and will therefore have sex appeal to the next generation of females. This has become known as the "sexy son hypothesis," and it is based on the intriguing notion that such choosy females will ultimately be more fit because their sexy sons will themselves be chosen preferentially, thereby bringing their choosy mothers a higher reproductive return by bequeathing them more grandchildren. Idea number two is that ornamented males (not just peacocks with their fancy tail feathers but a wide range of male traits, such as elaborate branching antlers, pendulous combs and wattles, deep voices, big tusks, etc., not to mention the often grotesque courtship displays that follow) are themselves biologically costly, so much so that only the best males can afford to produce them and still function effectively in the rest of their life activities.[38]

In short, such elaborate male traits are actually handicaps, and females choose males who do well in spite of them! After all, talk is cheap. It would be easy for a male to proclaim his superiority as a potential gene donor, but according to the "handicap principle"—which seems somewhat perverse when first considered—females have been obtaining good quality genes by choosing males who demonstrate their quality by encumbering themselves and then succeeding nonetheless.

These two hypotheses, sexy sons and handicaps, are not mutually exclusive. Indeed, it seems likely that males will be sexy precisely because they are able to rise above their handicaps. In addition, such sexily handicapped sons might very well also prosper in male–male competition, insofar as they show off the underlying capability of the male in question. One can imagine nearby peacocks saying to themselves "Wow! Look at that guy. He must be quite a tough customer—free of parasites and diseases, well endowed genetically—to be able to build and then tote around such a magnificent and yet basically useless tail!" And of course, peahens would be comparably impressed. (Which, in turn, would make that tail useful after all.)

The sexy son hypothesis could well apply to human beings, too. Just as a man doesn't need to make conscious assessment that a woman who is young, healthy, well nourished but not obese, sporting clear skin, healthy hair, adequate breasts, and full lips and hips is sexy because she is likely to bear healthy children, a woman needn't literally say to herself "I'd love to bear that guy's kids" when she finds him "cute," "sexy," or otherwise attractive. When it comes to our behavior, evolution works in many ways, most often through unconscious preferences.

It is at least possible that the "sexy son/handicap hypotheses," or some combination of the two, have been working among people, not just with regard to anatomy but behavior too. It is both debatable and controversial whether women are attracted to assertive, sometimes aggressive, even violent men. I would like to think otherwise, but there seems little doubt that especially during our long-duration prehistory, aggressive men would have been able—if nothing else—to provide protection for their mates against other, aggressive and perhaps nastier men.

The annals of evolutionary biology are full of examples whereby a trait is advantageous in moderate amount but disadvantageous if extreme. This is why most traits are present in a population in a "normal" or bell-shaped distribution: relatively rare at one end, then becoming increasingly abundant as the trait itself is more pronounced, and then progressively rare once again as the trait is increasingly abundant or pronounced. This distribution is found, for example, in birth weights, with around 7 lbs. 6 oz. being most common, and both lighter and heavier infants increasingly rare as the weight increases or decreases, a kind of thermostatic or negative feedback process with regard to maximum fitness.

Something like this doubtless operates with regard to male aggressiveness. It is a liability to be a total wimp, or on the other hand, to be so aggressive that one is constantly getting into confrontations. Women, too, are unlikely to be attracted to either a wimp or a hyperaggressive jerk, ultimately because such people are disadvantageous not only to themselves but to their immediate family. But male aggressiveness, and perhaps even violence, may well have been subject to a paradoxical kind of feedback, mediated at least in part by women's preferences. Thus, it is at least possible that increased male aggressiveness—as it became extreme and abundant in the local population—led not to selection for less aggressiveness but for more yet, as a result of female preference for males who could defend them and their progeny from other, aggressive males.

It is currently impossible to know how important rape has been in the human evolutionary past, but its threat may have selected a female preference for males who were capable of defending against this outrage—in the process, giving rise to sons who may themselves have had a low threshold for precisely the kind of sexual violence that is barely if at all distinguishable from the sexual jealousy we have already discussed.

Here is one "theory," expressed by Sylvia Plath in a selection from her poem "Daddy" that I fear captures some of the truth:

> Every woman adores a Fascist,
> The boot in the face, the brute
> Brute heart of a brute like you.

For some women, an abusive man may project an image of such power that the price of abuse is less than the payoff in being affiliated with him. In addition—shades of the sexy son hypothesis—brutish men would have been more likely to produce brutish sons, who, in turn, would enjoy the same advantages as their fathers. If so, then mothers who made such choices would give birth to sons more likely to survive and possibly be attractive to at least some women and who would, as a result, produce lots of (brutish) offspring. In other words, it is at least possible that independent of direct male–male competition, sexual selection operating via female choice may have produced violence and brutality among men in the same way that it has produced exotic and often bizarre traits among other male animals.

All this is debatable. Feminist anthropologists Adrienne Zihlman and Nancy Tanner, for example, have argued the opposite: that throughout our evolutionary history, women preferred men who were basically decent and cooperative:

> Females preferred to associate and have sex with males exhibiting friendly behavior, rather than those who were comparatively disruptive, a danger to themselves or offspring. The picture then is one of bipedal, tool-using, food-sharing, and sociable mothers choosing to copulate with males also possessing these traits.[39]

Maybe our ancestors actually preferred kinder, gentler men. Certainly, many women do today; perhaps most. The cross-cultural research by evolutionary psychologist David Buss shows that "Both sexes want mates who are kind, understanding, intelligent, healthy and dependable." However, as we have seen, male–female differences are also detectable, notably with regard to the valuing of physical traits (by men) and "resource-acquiring" traits (by women). Maybe some women, in addition, were turned on by violent SOBs. My guess is that most had to make the best of whatever—and whomever—came their way.

When describing evidence for human polygyny, I emphasized how harem defense confers a payoff to males who are comparatively large and often violent (sexual dimorphism), just as the previous discussion suggests that to some degree, this dimorphism might have arrived as a result of female choice. It is also possible, however—although I am not aware of anyone who has pursued this hypothesis at present—that sexual dimorphism arose as a consequence of polygyny via a different mechanism: not so much that males were selected to be larger but that females were selected to be smaller. Here's how it might have worked: under a polygynous mating regime, women who became sexually mature at a younger age would have more opportunity to join a harem. Because earlier sexual maturity correlates with smaller body size, selection for early menarche would have also selected for smaller females and thus, more sexual dimorphism. Why wouldn't a similar pattern have also selected for early sexual maturation of men? Because as already described, male–male competition would place early-maturing boys-into-men at a substantial and possibly violent disadvantage, one not shared by early-maturing girls-into-women.

There is in any event a regrettable association between sex and aggression, which needs to be confronted, as well as a gender-specific connection between sex and fear. Among males of many species, aggressiveness is fully consistent with sexual intercourse, while fear is inhibitory. It is very difficult for a fearful man to maintain an erection, whereas it is no coincidence that "arousal" can imply aggressive as well as sexual motivation. By contrast, fear does not inhibit sexual intercourse among female animals (although it may well inhibit female sexual gratification), and the tragic reality of human rape demonstrates not only that male sexuality coexists comfortably with aggression and even violence, but that fear does not prevent women from being victimized, and impregnated, by a determined—and altogether unwanted—sexual predator.

In Western culture, at least, it is clear that male-initiated sexuality is closely tied to aggression and occasionally violence. Even so nonbiological an observer as the humanistic psychotherapist Abraham Maslow pointed out that male sex is widely associated with manipulation, dominance, aggression, and physical control, as well as subordination of the erstwhile partner.[40] "Fuck you" is aggressive, not loving; and to be "fucked," "shafted," "screwed," and so forth is to be taken advantage of. Among many nonhuman primates, an erect phallus is used as a conspicuous threat display. Similar images and statues are widespread in the iconography of traditional human societies as well as in prehistoric art.

"Many men appear to take sexual pleasure from nearly all forms of violence," writes Susan Griffin. "Whatever the motivation, male sexuality and violence in our culture seem to be inseparable. James Bond alternately whips out his revolver and his cock, and though there is no known connection between the skills of gun-fighting and love-making, pacifism seems suspiciously effeminate."[41] There is, it should be clear at this point, a "known connection" between violence and male sex, a connection that arrives and/or is reinforced by polygyny.

It is thus likely that the evolutionary payoffs to successful, harem-inclined men would have come about in at least two different ways, both converging in consequence. On one hand, men, not unlike bull elephant seals, must compete (or at least feel that they must compete) directly and often violently with other men. And on the other hand, successful competitors are often more attractive to women, simply by virtue of their success or the prospect of their being successful in such competition, whether offensive or defensive. Such attractiveness might very well also have been accelerated by a kind of sexy son phenomenon. (It is also possible that a handicap process was operating as well, although this seems less likely.)

There are a number of examples from the biological world in which females test their prospective mates preliminary to mating as an important part of courtship. In the cactus mouse (*Peromyscus eremicus*), the male copulates several times with the female but ejaculates only after the first copulation. Subsequent mountings might seem to be nonfunctional because no sperm is transferred. However, the female cactus mouse only ovulates after these repeated "dry" matings,[42] apparently because only a male cactus mouse who is socially dominant is able to perform these sequential copulations; if he is interrupted, then he is revealed as a subordinate. And so, only after the male has proven his mettle by monopolizing the female for a prolonged period will the female proceed to invest her ova and the pregnancy and lactation that follows.

This is not to claim that human beings are the same as cactus mice. It is to claim, however, that there is much in our sociosexual lives that reflects basic male–female traits, which themselves have arisen in response to polygamy—while also enhancing it.

So far, we have looked at the basic parameters of human polygamy (chapter 1), its impact on violence (chapter 2), and on sexual preferences (this chapter). When it comes to sex appeal, it makes sense not only that men compete for access to women—especially women who offer at least the shadow of potential reproductive returns—but also that women prefer men who offer a reciprocal prospect of contributing to the fitness of each woman making such a choice. After all, our species is unusual among mammals, and even among primates, in the extent to which infants are born helpless and remain needy for years, even decades. In the next chapter, we consider how all this appealing, preferring, yearning, competing, desiring, seeking, attracting, repelling, hassling, and hoping plays out more directly when it comes to parenting. And we shall see that once again polygamy looms large.

Notes

1. Kinsey, A. C., Pomeroy, W. B., Martin, C. E., & Gebhard, P. H. (1953). *Sexual behavior in the human female*. Philadelphia: W. B. Saunders.
2. Platek, S., & Shackelford, T. K. (Eds.). (2006). *Female infidelity and paternal uncertainty*. Cambridge, England: Cambridge University Press.
3. For readers younger than 40 or so, this term ("carbon copies") refers to a common technique—now antiquated—of making copies before the advent of copying machines and computers. I for one don't miss it.
4. Whitfield, J. (2004). Everything you always wanted to know about sexes. *PLoS Biology, 2*(6), e183.
5. Even though it is increasingly clear that women are indeed more discriminatory than men, we now know that Bateman was off target when he described women's sexual style as "passivity."
6. Newport, K. G. C. (2006). *The Branch Davidians of Waco: The history and beliefs of an apocalyptic sect*. New York: Oxford University Press.
7. Garcia, H. A. (2015). *Alpha God: The psychology of religious violence and oppression*. Amherst, NY: Prometheus Books.
8. Pizzari, T., Cornwallis, C., Køvlie, H., Jakobsson, S., & Birkhead, T. R. (2003). Sophisticated sperm allocation in male fowl. *Nature, 426*(6962), 70–74.
9. Pound, N. (2002). Male interest in visual cues of sperm competition risk. *Evolution and Human Behavior, 23*, 443–466.
10. Geher, G., & Kaufman, S. B. (2013). *Mating intelligence unleashed: The role of the mind in sex, dating and love*. New York: Oxford University Press.
11. Kennair, L. E. O., Schmitt, D. P., Fjeldavli, Y. L., & Harlem, S. K. (2009). Sex differences in sexual desires and attitudes in Norwegian samples. *Interpersona, 3*(Supp. 1), 1–32.
12. Marlow, F. W. (2004). Male preferences among Hadza hunter-gatherers. *Human Nature, 4*, 365–376.
13. Escasa, M., Gray, P. B., & Patton, J. Q. (2010). Male traits associated with attractiveness in Conambo, Ecuador. *Evolution and Human Behavior, 31*, 193–200.
14. Buss, D. (1989). Sex differences in human mate preferences: Evolutionary hypotheses tested in 37 cultures. *Behavioral and Brain Sciences, 12*, 1–49.
15. Bokek-Cohen, Y., Peres, Y., & Kanazawa, S. (2008). Rational choice and evolutionary psychology as explanations for mate selectivity. *Journal of Social, Evolutionary, and Cultural Psychology, 2*, 42.
16. Townsend, J. M. (1995). Sex without emotional involvement: An evolutionary interpretation of sex differences. *Archive of Sexual Behavior, 24*, 171–204.
17. Gustavsson, L., & Johnsson, J. (2008). Mixed support for sexual selection theories of mate preferences in the Swedish population. *Evolutionary Psychology, 6*, 575–585.
18. Jonason, P. K., Li, N. P., & Madson, L. (2012). It is not all about the Benjamins: Understanding preferences for mates with resources. *Personality and Individual Differences, 52*, 306–310.
19. Malinowski, B. (1929). *The sexual life of savages in north-western Melanesia*. London: Routledge.
20. Pérusse, D. (1993). Cultural and reproductive success in industrial societies: Testing the relationship at proximate and ultimate levels. *Behavioral and Brain Sciences 16*, 267–322.
21. Orwell, G. (1933). *Down and out in Paris and London*. London: Victor Gollancz.

22. Pérusse, D. (1994). Mate choice in modern societies: Testing evolutionary hypotheses with behavioral data. *Human Nature, 5,* 255–278; Kanazawa, S. (2003). Can evolutionary psychology explain reproductive behavior in the contemporary United States? *Sociological Quarterly, 44,* 291–302.

23. Barash, D. P. (1981). *The whisperings within: Evolution and the origins of human nature.* New York: Penguin.

24. Miller, G., Tybur, J. M., & Jordan, B. D. (2007). Ovulatory cycle effects on tip earnings by lap dancers: Economic evidence for human estrus? *Evolution and Human Behavior, 28,* 375–381.

25. Tracy, J. L., & Beall, A. T. (2014). The impact of weather on women's tendency to wear red or pink when at high risk for conception. *PLoS One, 9*(2), e88852; Beall, A. T., & Tracy, J. L. (2013). Women are more likely to wear red or pink at peak fertility. *Psychological Science, 24,* 1837–1841.

26. Thornhill, R., & Gangestad, S. W. (2008). *The evolutionary biology of human female sexuality.* New York: Oxford University Press.

27. Barash, D., & Lipton, J. E. (2010). *How women got their curves, and other just-so stories.* New York: Columbia University Press.

28. Hrdy, S. B. (2000). *Mother nature.* New York: Ballantine.

29. Lévi-Strauss, C. (1969). *The elementary structures of kinship.* Oxford, England: Alden Press.

30. In a rather subtle permutation on the double standard, only rarely is childlessness attributed to "failure" on the part of the husband.

31. Ford, C. S. (1951). Control of contraception in cross-cultural perspective. *Annals of the New York Academy of Sciences, 54,* 763–768.

32. Wrangham, R. W. (1993). The evolution of sexuality in chimpanzees and bonobos. *Human Nature, 4,* 47–79.

33. Stolting, K. N., & Wilson, A. B. (2007). Male pregnancy in seahorses and pipefish: Beyond the mammalian model. *Bioessays, 29,* 884–896.

34. Anderson, M. J., & Dixson, A. F. (2005). Sperm competition and the evolution of sperm midpiece volume in mammals. *Journal of the Zoological Society of London, 267,* 135–142.

35. Nunn, C. L., Gittleman, J. L., & Antonovics, J. (2000). Promiscuity and the primate immune system. *Science, 290,* 1168–1170.

36. Gallup, Jr., G. G., Burch, R. L., Zappieri, M. L., Parvez, R. A., Stockwell, M. L., & Davis, J. A. (2003). The human penis as a semen displacement device. *Evolution and Human Behavior, 24,* 277–289.

37. Kenrick, D. T., Sadalla, E. K., Groth, G., & Trost, M. R. (1990). Evolution, traits, and the stages of human courtship: Qualifying the parental investment model. *Journal of Personality, 58,* 97–116.

38. Zahavi, A., & Zahavi, A. (1987). *The handicap principle: A missing piece of Darwin's puzzle.* London: Oxford University Press.

39. Tanner, N., & Zihlman, A. (1976). Women in evolution. Part I: Innovation and selection in human origins. *Signs, 1,* 585–608.

40. Maslow, A. H., Rand, H., & Newman, S. (1960). Some parallels between sexual and dominance behavior of infra-human primates and the fantasies of patients in psychotherapy. *Journal of Nervous and Mental Diseases, 131,* 202–212.

41. Griffin, S. (1971). Rape: the all-American crime. *Ramparts, 9,* 24–35.

42. Dewsbury, D. A., Estep, D. Q., & Lanier, D. L. (1977). Estrous cycles of nine species of muroid rodents. *Journal of Mammalogy, 58,* 89–92.

4

Parenting

Here is what anthropologists call a "cross-cultural universal": there is no society in which men do more fathering than women do mothering, in fact, none in which men and women are even comparable. As already noted, this is an unavoidable result of internal fertilization. Through evolutionary time, men simply haven't had a way to know for certain that their offspring are in fact theirs, to match the guarantee that evolution naturally provides to women. Taken by itself, this provides powerful biological momentum to the widespread social prohibitions against overt polyandry because enhanced paternal uncertainty correlates not only with male-generated violence but also with male disinvestment in the offspring. This italicizes why polyandry is so rare as an institutionalized human phenomenon and is nearly always covert (in animals as well as in people).

The key difference between men and women when it comes to confidence[1] of genetic relatedness to their offspring also explains the paradox that women, not men, lactate. The fact that lactation is done by mothers and not fathers seems obvious ... until we step back and ask why. After all, immediately after childbirth, when women have undergone the stresses of pregnancy and the birth process, it would seem only fair—and even fitness-enhancing—if fathers would pitch in and nourish their infants. But this isn't true for any people, anywhere, nor for *any* species of mammal. We all know about deadbeat dads; just a whiff of evolutionary insight makes it clear why there are so very few deadbeat moms. It also helps us understand why paternal uninvolvement is especially pronounced among polygynous species, and the more polygynous, the more deadbeat are the dads.

There is an ancient Roman saying: *pater semper incertus* ("the father is always uncertain"). And earlier yet, in Homer's *Odyssey*, we overhear Telemachus talking with Athena about Odysseus:

> My mother saith that he is my father;
> For myself I know it not,
> For no man knoweth who hath begotten him.

Of course, this unknowing is more potent yet when it comes to men's uncertainty as to who they have begotten, an issue with real consequences when it comes to what biologists call "parental investment." When a man is asked if he has any children, there is some comedy in the reply "Not that I know of"; but there is also a profound biological truth.

For males of most species—human beings included—substantial effort and often risk must be expended to obtain even one copulation; but having done so, the production of offspring can occur at relatively little further cost. Moreover, because male mammals lack the females' confidence in their genetic parenthood, they aren't adapted to nourish those offspring, whereas females are. As a result, males spend considerable time and energy at what is termed "mating effort," which often involves attempting to copulate with additional females rather than investing in their own— rather, in their mate's offspring, who may or may not be their own. On the other hand, females—especially in polygynous species—typically have little difficulty obtaining sex, and accordingly they spend more time and energy in "parental effort," investing in offspring that are reliably theirs.

As a general rule, reproductive success among female primates doesn't depend greatly on social rank. One review of nonhuman primates found that just slightly over one-third of the most dominant females (34%) produced barely more than their share (38%) of all offspring.[2] Comparable data aren't available for males of most species, although it is easy to predict that the situation is dramatically more skewed. Almost certainly, females enhance their overall fitness by enhancing their social status, but less so than is the case for males. What females *can* do to maximize their fitness, on the other hand, is to focus on their parenting. For females, the emphasis, not surprisingly, is on quality; whereas for males, it's the other way around: the more mating (i.e., quantity), the better.

Obtaining additional matings, however, isn't all that easy—if you're male. Even when violence isn't at issue, male mating effort is notoriously more risky than its parental counterpart. This contrast has been nicely described by Martin Daly, the evolutionary psychologist whose work on homicide we encountered in chapter 2. Before turning his attention—along with that of his wife, Margo Wilson—to *Homo sapiens,* Daly studied the behavior and ecology of free-living gerbils. "In the Sahara Desert," he writes,[3]

> live several species of gerbils, small rodents much like the popular pet gerbil that originally comes from Mongolia. Most gerbils are rather solitary animals. Wherever a little plant life can be found in the harsh Saharan terrain, individual females inhabit and defend small domains adequate to provide for themselves and their pups [*parental effort*—DB]. The distribution of females is thus related to food resources, but the critical resource for males is the females themselves. The imperative of male gerbil existence is getting around to call on the females [*mating effort*—DB]. Males strive to visit as many females as possible and as often as is necessary to catch each female on her day of sexual receptivity. Toward this end males are constantly making their rounds. They are apt to inhabit burrow sites that are too poor in food ever to appeal to a female but that are centrally located for visits to several females living in scattered food-rich areas.

These gerbils afford another example of the perils of the competitive male life-style. The female has a very good chance of living a year or more; she sticks to a small, well-known area with lots of cover and she maintains several minor escape burrows. A male, on the other hand, repeatedly takes his life in his paws by running hundreds of meters across open country in order to visit a female who is certain to drive him away, but who may just possibly be willing to mate with him before doing so. It is a dangerous quest. The entire adult male population of an area is likely to turn over every few months, while females rest secure in their territories. Despite equal numbers of each sex at weaning, adult female gerbils are sometimes two to three times as numerous as males.

Earlier, we reviewed and summarized the connection between polygyny and sexual dimorphism: in which males are significantly larger than females. As with so much in biology, however, things aren't nearly so cut and dried. There are a number of mammal species in which things are reversed, with females larger than males, and these aren't cases of polyandry (polyandrous mammals are even rarer than polyandrous birds, which is saying quite a lot). Females are larger than males among such species as chinchillas, cottontail rabbits, duikers (tiny African antelopes), marmosets (small New World monkeys), and more bats than you can fit into the average belfry.

One possible explanation—suggested most strongly by biologist Katherine Ralls[4] several decades ago—has become known as the Big Mother hypothesis: given that pregnancy and lactation are demanding, it makes sense that a larger, heavier, sturdier mother would likely be a more successful one. Remember sexual bimaturism in which males become socially and sexually mature later than females? Given the greater physiological demands that the biology of reproduction makes on mothers rather than fathers, this pattern emerges as part of the suite of characters that speak "polygyny" to any attentive biologist.

We simply don't know why females of some species are larger than males, and it is at least possible that Ralls's Big Mother hypothesis captures at least some of the reason, which, as already noted, doesn't include polyandry. The Big Mother hypothesis makes sense especially for bats among whom pregnant females have to fly around carrying their unborn fetuses, presumably rendering an extra dose of muscle power more fitness-enhancing than would being especially svelte. Thus, it isn't always true that mating effort (a male specialty) trumps parental effort (a female specialty) when it comes to selection for body size. In any event, it is significant that none of the mammals in which females are bigger than males are highly polygynous.

It is also important to appreciate just how much female mammals invest when it comes to parental effort. Let's go back to those elephant seals that we met in chapter 1. A female who weighs perhaps 700 kg gives birth to a pup weighing about 50 kg. That's 1/14 of her weight, equivalent to a 140 lb. woman producing a 10 lb. child, and thus not initially that different from a human mother's beginning parental effort. But then, in the next month or so, the elephant seal pup will roughly double its birth weight, going from 50 to 100 kg, all of it due to having consumed its mother's milk, and during which the mother will lose about 2 kg for every 1 kg conveyed to her pup. During this time, the pup's father hasn't expended *any* of his roughly 2,500 kg on behalf of the pup, nor for any of the other pups (as many as 30 to 40) that he also sired. He's all about mating, about becoming a father but not about being one.

Despite the focus of male mammals on mating effort, some species, particularly those whose polygyny is more moderate, show definite paternal behavior. And this includes *Homo sapiens*. Although human parenting is nonetheless primarily "women's work," men are obviously capable of doing a lot of fathering, perhaps more than any other species. A detailed meta-analysis of the reproductive ecology of 186 human societies in which there was substantial paternal investment found that provisioning offspring was a universal trait that resulted in significantly greater survival for the father's children.[5]

Given that human beings require a huge amount of parental investment, it makes sense that despite our polygamous heritage, people are also predisposed toward biparental care. For men, the conflicting pushes and pulls of monogamy versus polygyny, parenting versus gallivanting, and so forth reflect biology interacting with cultural traditions, personal preferences overlain on social responsibility. By the same token, women experience conflicting impulses toward both monogamy and polyandry, toward experiencing a variety of sexual partners yet also assuring biparental care of any offspring thereby produced, all the while negotiating the often stringent demands of diverse (and changing) cultural traditions.

Regardless of the specific social group or its environmental situation, women worldwide typically work about as much—which is to say, very much indeed—at parenting. Men, by contrast, are considerably more variable in how much parental care they contribute. (Often, the bottom line is: as little as possible.) Even in modern Western societies in which women have made substantial progress in their representation in the workforce, they do substantially more work at home than do their husbands.[6]

It is very difficult—most likely, impossible—to come up with an animal species that serves as a useful biological model for parental behavior in our own species. And so, it is far easier—and almost certainly, more instructive—to identify species that are *not* suitable as models and to emphasize why they are not. We are not like African hornbills, for example, birds in which the female incubates her eggs in a hollowed-out tree cavity, after the male has literally walled her inside using daubs of mud, until just a narrow slit remains, through which he dutifully feeds her (and which is too narrow for predators, such as snakes, to enter). The male hornbill is a reliable provider and thus, in a sense, a doting father—but not directly toward his offspring.

We are also unlike either bonobos or chimpanzees, species in which the closest approximation to paternal care is the rather minimal contribution that among chimps, males refrain from killing infants if they have previously copulated with their mothers. Nor are we like barnacles in which male and female anonymously squirt their gametes into the ocean and no one does any parenting. Men, by contrast, are often quite paternal, even capable of rearing offspring as a single parent (albeit so long as someone else provides the milk, or milk substitute). Fathers matter, not just as someone to pace the sidelines during Little League games or occasionally chauffeur a child to the orthodontist. The data are quite convincing that among traditional peoples in particular, presence of a biological father contributes substantially to the success—even the survival—of children.[7]

The importance of *parenting* goes beyond the importance of *parents*, the simple fact that any sexually reproducing species needs a union of male and female gametes to reproduce. There is no doubt that the human need for parenting comes from the helplessness of our infants and how much time, effort, protection, and teaching they need, not just to survive but to thrive, which in evolutionary terms largely means to become successfully reproducing adults in their own right. This simple, crucial fact—our primordial infant helplessness combined with our prolonged and vulnerable childhood—has implications for that other simple, crucial fact: our polygamous inclinations.

First, let's consider the polygyny side of these inclinations and how they interface with parenting. It stands to reason that a harem-dwelling co-wife will have less attention and assistance from her husband (and presumed father of her children) than would a monogamously mated woman. But this doesn't necessarily mean that a co-wife will be less well off—in evolutionary terms, less fit—than her monogamous peers. In an appendix to his play, *Man and Superman*, George Bernard Shaw wrote that "the maternal instinct leads a woman to prefer a tenth share in a first rate man to the exclusive possession of a third rate one."

In short, the Shavian wisdom—asserted by evolutionary biologists decades later—is that many women would be better off married to a multimillionaire, even if they had to share him and his wealth with other co-wives, than to be monogamously mated to a loser—or even, to an average provider. As we'll see, evidence from traditional societies strongly suggests that this is probably *not* the case; that is, *contra* Mr. Shaw, polygyny really does seem to be a bad deal for women. (Just as its competitive consequences make it a bad deal for men, too—at least, for most men.) But for now, let's pursue the more traditional biological interpretation, which conforms rather closely to Shaw's insight.

Nearly half a century ago, ecologist Gordon Orians and his graduate student Jared Verner developed an approach to understanding the evolution of polygyny that has largely withstood the elegant permutations of emerging theory as well as the predictable efflorescence of empirical findings.[8] Known as the "polygyny threshold model," it was initially applied to birds but has subsequently been found generally applicable to mammals and, with only slight modification, to *Homo sapiens* as well.[9]

It is early spring in a cattail marsh in North America regularly occupied by red-winged blackbirds. Male red-wings are just returning from their southern migration. They sing, display to one another, and occasionally tussle, eventually establishing territories on the marsh. Shortly afterward, the females begin to arrive, initially distributing themselves across these territories and their male proprietors. But soon, something interesting happens. Imagine that the marsh is divided into 10 territories, each occupied by a different male. After perhaps six females have arrived, each settling into a different territory, female number seven shows up and instead of affiliating with male number seven, she sets up housekeeping with male number one, who already has a mate, thus becoming part of his "harem" of two.

As the migration concludes, a total of 10 females has arrived at the marsh, but instead of this yielding 10 monogamous pairs, there may be one male with three females, two more males with two females each, three males with one female each, and four males with no lady red-wings at all. As described in chapter 1, the males in question will almost certainly experience a higher variance in reproductive success than will the females because polygyny has popped up, along with a dose of monogamy. But some questions arise.

A particular strength of the polygyny threshold model is that it examines this situation from the perspective of the females, asking this question: why have some of them chosen to be mated polygynously—to join a male's harem, as a result of which they only get "part" of that male's parental assistance—when they could have been the sole mate (maybe even the soul mate) of a seemingly eligible territory-owning bachelor, now left to languish unmated?

The answer, leaving behind the mathematics and nifty graphs, is simple. Granted there is presumably a payoff to a female if she chooses to be monogamously mated because she and her offspring get the full-time attention of her "husband." Nevertheless, being part of a harem could be an even better option for her, if an already-mated male is "rich" enough (which is to say, if he offers enough resources) to make up for the loss of his undiluted attention. The polygyny threshold, in this model, is the difference in resources between two competing males—one a bachelor and the other already mated—to make it worthwhile for a female to choose the latter.

The underlying assumption is that from the female's viewpoint, for polygyny to be selected[10] there must be enough of a difference in what competing males have to offer. In a zoological context, this leads to the prediction that polygyny would occur when there is sufficient heterogeneity in local habitats (suitable nest sites, food abundance, and so forth) and also among species in which paternal care is relatively unimportant (thus, "precocial" species such as hoofed mammals). Also, of course, females must have the option of exercising choice.

The polygyny threshold model does in fact apply quite well for many animals, especially birds that are precocious and that occupy heterogeneous environments. The nature of male "wealth" varies with the ecology of each species, including such considerations as availability of predator-safe, elevated nest sites in European pied flycatchers;[11] the density of cattails (again, an antipredator advantage) in North American red-winged blackbirds;[12] and the availability of shaded nest sites (providing relief from summertime heat stress in the open prairie) among lark buntings.[13]

The model does a good job of describing the situation among at least some mammals, too. Gunnison's prairie dogs, for example, are somewhat monogamous when the richness of their prairie grass food source is homogeneous across their habitat; they become polygynous, however, when the distribution of their preferred groceries is patchy.[14] Male prairie dogs who are "wealthy"—who occupy rich habitats—mate with multiple females. Even though polygynous males do well in the animal world, it isn't necessarily so that females are similarly benefitted. For example, among yellowbellied marmots—western relatives of the eastern woodchuck, in which males can occupy themselves with one, two, or three females—the reproductive success per female decreases as harem size increases.[15]

What about human beings? The polygyny threshold model comports quite nicely with Shaw's quip, wealthy men replacing red-winged blackbirds who occupy an especially fitness-promoting bit of avian real estate. We would expect polygyny to be more frequent and more extreme in situations of greater income and status inequality (the equivalent of habitat heterogeneity), and also that powerful, wealthy men should be more likely to end up polygynously mated. As we've already seen, this is definitely the case, because despotism and income inequality correlates with harem size. As previously noted, the Qur'an explicitly states that a man can have up to four wives, with the number determined by his wealth: whether or not he is capable of supporting them and their children. "Marry women of your choice," we find in sura 4:3, "two or three or four; but if ye fear that ye shall not be able to deal justly [with them], then only one."

Positive associations between resource control and male reproductive success have been found for numerous traditional societies, including but not limited to Yomut Turkmen;[16] Trinidadians;[17] inhabitants of the tiny Pacific island of Ifaluk;[18] as well as among the !Kung, Murngin, Nuer, Taloensi, Rajput, and mainland Chinese.[19] In fact, this correlation has been found so consistently that most researchers these days don't even bother to evaluate it (another one of those "Duh!" phenomena, like women's preference for high-achieving, resource-rich men, and men's preference for young, physically attractive women.)

An outside possibility that has received insufficient attention is that polygyny occurs in some cases as a reflection of males' efforts to *avoid* parental investment (rather than being a choice by females in which male resource quality compensates for the loss of paternal help). There is at least one animal example of this kind of thing. Among tree swallows, some males are monogamous, maintaining traditional territories within which they mate and help rear their offspring; others are "floaters," who had been considered to be nonbreeders, relegated to their second-tier status because of their presumed poorer physical condition. Surprisingly, however, it was found that these floater males not only reproduced—often more successfully than their monogamous compatriots—but were generally in superior physical condition.[20] This leads to the possible conclusion that better males are able to mate polygynously—presumably because they are more attractive to already mated females than are the existing social mates of these females—and then to shrug off the demands of child care onto those dutiful monogamous males, who are essentially parasitized by these love-'em-and-leave-'em cads.

We have already noted that even in formally monogamous human societies, polygyny often occurs as well. Significantly, when it does, it is nearly always an achievement of men who can provide their mates with a greater-than-average share of resources. For example, the Kalahari Bushmen inhabit a stringently resource-limited environment and are primarily monogamous as a result; in such circumstances it is difficult for a man to accumulate substantially more of anything than his fellows. Among them, 5% of the men nonetheless manage to have two wives.[21] In this and other cases, male success is less influenced by aggressiveness or violence than by ability to obtain resources that, in turn, are appealing to women: for example, being a proficient hunter. In such cases, the proximate lure may well be the prospect of regular, nourishing meals, whereas the ultimate payoff is likely to be reproductive.

If a small number of men have monopolized all of the relevant resources, then there is no question that females are "better off" mating polygynously, simply because if they tried to breed monogamously, they'd get none at all! Along these lines, primatologist Sarah Hrdy quotes 19th century Darwinian feminist Eliza Burt to the effect that "the female of the human species is obliged to captivate the male in order to secure her support."[22] Moreover, if there is a major threat from marauding bands of inimical human beings or threatening predators, a woman would doubtless be more fit if associated with a powerful and reliable protector. In such cases, the way to obtain security as well as access to resources is—at the risk of sounding cold-hearted about it—to barter one's eggs and fertility in return.

In other cases, it's a trade-off. The unfortunate truth is that most women in most polygynous human societies don't have much choice; typically, the decision is made by their parents and other relatives, including notably their older siblings. As to whether they are better off under polygyny, the data are mixed. We've already noted the abundant evidence that—as predicted by the polygyny threshold model—wealthier men have more wives, and thus, more children. But another prediction—that polygynously mated women would on average do at least as well as their monogamously mated peers—is generally not fulfilled.

In some cases, it seems clear that women are worse off by virtue of polygyny. A particularly well-studied example are the Dogon people of Mali, in West Africa. Much of what we know about these people—and it is quite a lot—we owe to anthropologist Beverly Strassman.[23] Dogon society is very male structured: wives move to their husband's residence (patrilocal); wealth is inherited in the male line only (patrilineal); and men control the social, economic, and political levers of power (patriarchal).

When a Dogon woman menstruates, she is required to move to a special hut. By this means, her fertility is closely monitored. This fertility—as well as her sexuality—is further constrained by female circumcision, whereby her clitoris is removed, making sex much less enjoyable and presumably diminishing a Dogon woman's inclination to risk infidelity.[24] Sarah Hrdy notes that "In this way, older men with several young wives can be as certain of paternity as any primates in the world—Dogon certain."[25]

Despite all this, it turns out that Dogon husbands aren't very dutiful fathers, spending most of their time and energy in their mating efforts. They seek to enhance and maintain their social status, and with it, opportunities for more wives and thus more children to be ignored. And so, child mortality is terribly high, with nearly one-half dying before reaching five years of age. Moreover, a Dogon child living in a polygynous family is 7 to 11 times more likely to die early than if she or he is born to a monogamously mated woman. This appears to be due to the crucial reality that with monogamy, a woman's success is also her husband's, and ditto for losses. Hence, monogamously mated men are more likely to cooperate with their wives in maximizing the success of their shared offspring.

Polygynous males, on the other hand, are evolutionarily rewarded for emphasizing quantity over quality: a "father" with three wives, each of whom suffers about a 50% probability of losing her child, still ends up with 1.5 children on average compared to a monogamously mated man, who, even assuming zero mortality among his offspring, ends up with only 1.0.

It is undeniable that among the Dogon, at least, even though polygyny is a good deal biologically for harem-holding men, women aren't well served by it; nor are their offspring. The situation is complicated, however, because here—as with most polygynous societies—co-wives are not created equal. Nor are they treated equally. In particular, there is typically a disparity between the senior and junior wives. Nearly always, the senior wife is more powerful and thus better off under polygyny than are the younger, subsequent wives. It is therefore possible that at least some harem-dwelling women aren't necessarily worse off as the reigning male acquires additional mates.

Among the Mende people of Sierra Leone, for example, monogamously married women produced more children than did polygynous co-wives, *on average*. However, within these polygynous mateships, the most senior wives not only had greater reproductive success than their junior compatriots but more than monogamously mated women.[26] Because there is typically a dominance hierarchy among co-wives, it may only be the "junior wives" who suffer and who might do better if they were monogamously mated. At the same time, newer wives—because they are younger— are more fertile, albeit their offspring are more at risk because of their mothers' lower social status.

The institution of polygyny among the Dogon is worth investigating for several reasons, not least because their way of life has been rapidly disappearing. Beverly Strassman notes that in Dogon society, it is taken as a given that at marriage, the bride should be much younger than the groom.[27] This culturally generated ideology strongly favors the reproductive interests of wealthy, high status men. It is scarcely the only example of this kind of phenomenon. Cultural rules and expectations are biased, species-wide, in favor of men (powerful men in particular) and, not surprisingly, such men asymmetrically reap returns, both immediate—wealth, power, prestige—and evolutionary.

It isn't clear precisely why the children of Dogon co-wives suffer such high mortality, although it is widely claimed—among Dogon women themselves—that they are often poisoned by other co-wives. Young boys are notably vulnerable to such mortality; if they are directly targeted by others within the polygynously "distended family," it might be because daughters typically leave home when they marry outside the family—recall that the Dogon are both patrilocal and patrilineal—so that sons are left to compete among themselves for the father's inheritance. This, in turn, could well predispose especially vicious and self-serving Dogon mothers to eliminate their own sons' competitors.

There is something downright biblical about this scenario, and literally so: recall the actions of Sarah, depicted in the Bible. When she had been unable to conceive, Sarah "loaned" her handmaid, Hagar, to her husband, Abraham. The result was Ishmael, who incurred Sarah's resentment and competition relative to her own eventual son, Isaac. The upshot is that Sarah arranged for Hagar and Ishmael to be banished, an event that, as the story goes, was hugely consequential for subsequent human history: Ishmael was reputedly the progenitor of the Arabs, who—at least in the current Middle East—continue to compete with the descendants of Isaac, the Jews.

Beyond the dramatic finding that nonbiological parent figures comprise the greatest risk factor for children's neglect, abuse, and infanticide, there is evidence that simply living among nonbiological "parents" can be stressful, and not just among the Dogon. More than 15,000 saliva samples were collected from children living on the Caribbean Island of Dominica. This array of spit was then analyzed for cortisol level, a good indicator of stress. The finding? Children living with stepfathers, half siblings, and more distant relatives had abnormally high cortisol levels, indicating that they were living under substantial stress compared with their peers growing up in biologically intact, monogamous households.[28] Although none of the Dominican children were literally in polygynous situations, it is suggestive to extrapolate to such cases—that is, when nonbiological parents comprise an important part of the immediate household—in which children would likely be experiencing higher stress, even if they aren't literally neglected, abused, or murdered.

We might have expected Dogon co-wives to cooperate with each other so as to maximize the payoff to all, but that doesn't appear to happen, perhaps because among these people inheritable land is scarce; and given the custom of leaving all or nearly all of it to one or a very small number of sons, Dogon co-wives may be strongly tempted to promise cooperation, then defect lethally on occasion. And it only takes one such act to eliminate a competitor to one's own offspring.

Nor are the Dogon unique. Other social groups resemble them—and those yellow-bellied marmots—in that harem-dwelling females are worse off, reproductively, than are monogamists. Out of a sample of 246 married men among the Temne people of Sierra Leone, 133 were polygynists who were more "fit" (i.e., more surviving children per man) than were monogamous men. Not only that, but their "fitness" increased with corresponding increases in harem size. Temne women, however, were a different story. One quarter of the children born to monogamously mated women failed to survive infancy, whereas among bigamously mated women, infant mortality was a whopping 41%.[29]

Here is another example: among the Kipsigis people of Kenya, the number of surviving children declined with the number of co-wives. Thus, a monogamously married Kipsigis woman had on average 7.05 surviving children; a bigamously married woman, 6.82; whereas women with two or three other co-wives had 5.58 and 5.81, respectively.[30] In short, when a man decides to take a second wife, this is nearly always bad news for the first one. She will have to share resources and almost certainly compete with the new—and likely younger—wife and her eventual children.

There are other convincing data that polygynously mated women in sub-Saharan Africa suffer more than do those who are monogamously mated: more mental illness, especially depression, and more physical abuse.[31] In addition, their children fare worse by most measures.[32] The situation is typically most challenging for the most junior wives, who are more likely to have lower reproductive success because their infants suffer higher mortality rates.

Inclusive fitness theory (also known as kin selection) suggests that the liabilities of polygyny would be less when the co-wives are sisters, so-called "sororal polygyny." This is analogous to "polyandry" in which the co-husbands are brothers. Although, as noted earlier, polyandry is about as rare among animals as among people, when it occurs in the "animal world" it is likely to be "fraternal," with the co-husbands being brothers. In one well-studied case, that of the Tasmanian native hen (actually a species of flightless rail), females are most often monogamously mated, but occasionally mate with two males in rapid succession. When this is the case, the males are nearly always brothers.[33] From the perspective of the males involved, polyandry normally doesn't seem to have much to recommend itself, except that when the males are brothers, in the worst case—if a given male's sperm don't get to fertilize any eggs—each male is guaranteed to be at least an uncle, so his "inclusive fitness" is above zero.

Sororal polygyny should similarly benefit not only the co-wives but also the harem keeper, and for the same reason: there should be less conflict among the wives because each is an aunt to the other's children. Hence, although each sororal co-wife can be expected to favor her own offspring, she also has a genetic interest in the success of her nieces and nephews.

There haven't been many quantitative studies of reproduction in sororal harems. One of the most impressive comes from research on Australian aborigines living in southeast Arnhem Land.[34] Among these people, women who were part of a harem experienced fewer pregnancies, fewer live births, and higher mortality among their infants than did monogamously married women. These disadvantages were less, however, when the co-wives were sisters.

Also worth mentioning is the suggestion that in some cases—for example, among the Ibo of Nigeria[35]—monogamously mated women actually encourage their husbands to take an additional wife, the claim being that it is not only lonely to be an "only wife" but humiliating as well because it suggests that one's husband isn't very successful or otherwise desirable. It is also possible that even though Dogon women are reproductively ill served by polygyny (in terms of their success in rearing offspring), they recoup these losses via enhanced numbers of grandchildren if their sons are more likely to become successful polygynists: another case of greater reproductive success by "sexy sons." To my knowledge, this has not been investigated.

The evidence for polygyny being a reproductive disadvantage to women is far from a "slam dunk," for yet another reason. At least sometimes, men take additional wives because their first wife is infertile. The effect would be to reduce the total number of surviving children per wife in a polygynous household, but not necessarily because such arrangements are less productive for every woman involved.

Not surprisingly, however, when women are free to express their views, they are generally disapproving of polygyny. Indonesia is the most populous country in which polygyny is legal; roughly 5% of currently recorded marriages in that country are polygynous. Here, women often engage in street protests against polygyny. It seems clear that even when it is in the economic interest of women (assuming that to be the case), polygyny is not generally in their psychological interest—and as we have seen, only sometimes is it in their biological interest. Much of the time, women are coerced into polygyny, either because of direct pressure from parents and brothers (as well as their future husband) but also by the pressure exerted by a socioeconomic system that leaves them with very little wealth or power.

So, is polygyny beneficial to women, or not? It depends—mostly on the amount of freedom women have and on the quantity of wealth and other resources on offer. The upshot is a bit like the arguments surrounding prostitution. Some people—including some feminists—argue that women should have the right to sell their bodies as they wish in the marketplace, such that they are in effect taking advantage of the neediness and vulnerability of men who are forced to pay for something that most women can get for free. This could be valid for high-end "society girls" who make substantial incomes; but it is certainly not the case for the great majority of women who are grossly underpaid, not to mention physically and emotionally abused by a male-dominated system that clearly takes advantage of them. Polygyny is much the same.

No one doubts, however, that among polygynous societies, women are particularly likely to end up with men who are especially high status and wealthy. There remains, however, debate as to why. For example, it has been suggested that women "marry up"—whether this means joining a polygynous household or simply choosing to mate monogamously with men who are socioeconomically their "superiors"—simply because this is the best option they have for improving their social situation.[36] A good statement of this position appears in an article titled "The Origins of Sex Differences in Human Behavior: Evolved Dispositions or Social Roles?" and its conclusion: social roles.

This view is intuitively plausible. After all, there are no truly matriarchal societies in which women have more social and economic power than men, which—rather than evolutionary pressures—could also explain why men are much less prone to hypergamy. Women marry up and men marry down. And because men in polygynous situations are especially likely to be already "up" in that they control resources (plus the fact that women don't have much anyway), the impact of social roles on female hypergamy is consistent with the finding that men aren't nearly as concerned as are women about the wealth or status of their spouses.

Current evidence, however, does not support the social role or "structural powerlessness" theory. Numerous studies, investigating women's marriage preferences in many countries—including those as different as Jordan, Spain, and Serbia—conclude that higher status, wealthier women prefer men who are of higher status and greater wealth yet.[37] A parallel series of studies found that women who are judged to be especially attractive, either by others or by their own self-evaluation, also are inclined to prefer male partners who are especially attractive; in this case, attractiveness correlates with social and financial status, physical appearance, intelligence, and so forth.[38]

On the other hand, it is entirely possible that one reason polygyny is so much more stable than is polyandry is that in most societies men have more power than women and therefore they are able to leave a polyandrous system that frustrates them, whereas women are typically more dependent and powerless, hence, they are less able to leave an unsatisfying polygynous situation.

One of the most consistent natural history facts is the equal male:female sex ratio. It is found in nearly every animal species and is somewhat counterintuitive. In highly polygynous species, for example, wouldn't it be more efficient if the sex ratio matched the breeding system? If, say, 1 male typically mated with 10 females, then wouldn't it be appropriate for there to be 1 male produced for every 10 females? Instead, we find that regardless of the breeding system, there are almost exactly equal numbers of males and females; in a hypothetical example in which harem-keeping males have, on average, 10 mates, for every such successful male, there are 9 unsuccessful bachelors, making trouble and not contributing anything to the species breeding success. (Indeed, the reason they make trouble is that they are unable to breed and hence desperate.)

Biologists have a good explanation for the equal male:female sex ratio, even under extreme polygamy, starting with the fact that natural selection doesn't operate at the level of the species but is focused instead on the individual and his or her genes. From the perspective of potential parents, producing a daughter (who is pretty much guaranteed to breed) is precisely as good a bet as producing a son, even though in our example above, each son has only a 1 in 10 chance of success. The key is that this 1 successful male offspring will hit the jackpot, exactly balancing the failures of the 9 nonbreeding males and equaling the reproductive success of each female. Let's imagine that every female has 1 offspring per year, so the annualized fitness of a parent who produces a female is 1. What is the annualized fitness of a parent who produces a male? It is also 1: 9/10 of the time it is 0, but 1/10 of the time it is 10, for an average fitness payoff of, once again, 1. A similar result occurs regardless of the breeding system, whether monogamous, polygynous, or polyandrous. Not only are parents equally fit whether they produce boys or girls, but if the sex ratio deviates at any time (perhaps because of random differences in mortality), then selection will favor producing the sex that is less abundant, thus bringing the sex ratio back to 50:50.[39]

So far, so good: indeed, for evolutionary biologists, it is very good indeed because the convergence of theory and data in this analysis is very satisfying.

But it turns out that there is an important exception, an underlying assumption that isn't always met, namely, that mothers who are producing daughters or sons don't have any way of "knowing" whether their offspring can be predicted to be somewhat more, or somewhat less, successful than the average. If they have no relevant information, then as we have seen, making either a son or a daughter is an equally good evolutionary bet. Things change, however, if a prospective parent is somewhat healthier or more socially dominant than the average. In this case, any male offspring are somewhat more likely to be harem masters than chance alone would predict.[40] As explained in chapter 1, males in any polygynous species represent more high-risk bets than do females, simply because of the higher variance in male reproductive success. But parents who already have some sort of social or biological advantage can be expected to bias the sex ratio of their offspring to favor males because each male has better-than-average prospects. The inverse occurs for parents who are somewhat less healthy or less socially dominant: they would be more fit producing females, which, after all, become a safer bet because compared to males, their success is less dependent on competition with others.

This prediction has not been universally confirmed among animals, although it looks to be generally robust, occurring, for example, in animals as different as seals,[41] reindeer,[42] and dairy cows,[43] although evidence from nonhuman primates is equivocal.[44]

According to anthropologist Mildred Dickemann, it even tends to hold among human beings. Although there doesn't appear to be any effect of physical health or socioeconomic situation on the biologically generated sex ratio at birth, in highly stratified societies (in which polygyny is de facto and often de jure as well), upper-class families invest more in their sons and lower-class—as expected—in their daughters.[45] This makes a kind of evolutionary, reproductive sense, insofar as upper-class sons have the potential of producing many offspring, whereas upper-class daughters are, simply because of their womanhood, unlikely to have extraordinary numbers of children. And of course, the opposite applies to lower-class parents: their sons are likely to end up disappointed bachelors (or in any event, unlikely to be highly successful), whereas their daughters enjoy better prospects.

Interestingly, a study of children in Florida—published in 2015—found that boys fare worse than girls in low socioeconomic households. The authors conclude that under the same degree of social and economic disadvantage, boys are more negatively affected than are girls, a gender gap that is consistent with the greater biologically inspired need for boys to distinguish themselves, to establish themselves, to get ahead in a primordially competitive world.[46]

The linkage between sex and reproductive success leads to a tantalizing alternative—or additional—explanation for why men are bigger than women, one that plays on male–male competition under conditions of polygyny but adds another wrinkle. Insofar as larger parents produce larger offspring and smaller parents produce smaller offspring—which is certainly true, on average—then if heavier and taller people are more likely to produce boys, because larger sons are more likely to be reproductively successful, and lighter and shorter people are more likely to produce girls, because doing so is a more conservative bet under those conditions, this in itself would contribute to male–female sexual dimorphism. The available data suggest that this occurs among human beings.[47]

We have seen that where polygyny is possible, males in particular would benefit reproductively from control of resources because this would likely result in having multiple wives, or at least multiple matings. Another prediction derives from this: that under conditions of polygyny, parents would be biologically rewarded if they preferentially endowed their sons, rather than their daughters, with inherited wealth. No doubt, women with resources would also be more evolutionarily fit than those without; but once a threshold is reached, even a very wealthy and healthy young wife isn't likely to produce significantly more children than is a moderately wealthy one, whereas a very wealthy son could father dramatically more children as a result of his endowment. Hence, under such conditions, families who invested heavily in their sons would reap a payoff of more grandchildren.

This prediction that sons would generally receive more inheritance than daughters was tested by evolutionary anthropologist John Hartung, using two gold standards for anthropological research: G. P. Murdock's *Ethnographic Atlas*[48] (which presents data for 1,170 societies) and the "Standard Cross-Cultural Sample" (SCCS; data for 186 societies)[49]. It was strongly upheld in both cases.[50]

This preferential investment in sons, particularly on the part of upper-class families, isn't only reflected in patterns of inheritance. In the great majority of human societies, wives are essentially purchased by the groom's family, who, in doing so, are investing in their sons and thereby in their own eventual grandchildren. The general pattern is that young, healthy, and thus attractive women (who not coincidentally have better reproductive prospects) cost more than do potential wives who are older, less healthy, and therefore less desirable.[51] The result, of course, is that wealthy, powerful men have more wives, younger wives, and more children than do run-of-the-mill men. This is particularly true in traditional societies.

In the 21st-century West, most people think of dowry payments—the system in which the bride's family provides resources (whether money or some other contribution) to the groom as part of the marriage arrangement—as quaint and outmoded. But most people also assume that dowry payments are not only traditional but the most common pattern worldwide. Precisely the opposite is the case. Bride price, in which the groom's family pays the bride's, is found in 52% (97 out of 186) of societies listed in the Standard Cross-Cultural Sample and in 66% (839 out of 1,267) of societies found in the *Ethnographic Atlas*.[52] Dowry payment is a minority tradition, one that is primarily restricted to some East Asian societies and others largely from southern Europe.

Given that females of every mammal species are likely to be the reproductively limiting resource for males, it makes sense that in the case of human beings, bride price, not dowry payments, should be the default condition, with men (and/or their families) paying for reproductive access to women. Consider these irritating lines from Jonathan Swift's *A Gentle Echo on Woman*, politically incorrect, male chauvinist (avowedly so and intentionally provocative), as well as biologically perceptive:

"Say, what can keep her chaste whom I adore?"
—"A door."

"If music softens rocks, love tunes my lyre."
—"Liar."

"Then teach me, Echo, how shall I come by her?"
—"Buy her."

An evolutionary take on human mating systems should also deal with the seeming exceptions, raising the question why dowry payments occur at all. Given the polygyny threshold model, we would expect that particularly when some men control more than their fair share of reproductively relevant resources, women should cluster around such men. On the other hand, what about when monogamy is—for whatever reason—institutionalized? And even more so, what about when monogamy occurs along with substantial differences in the wealth of men?

In such cases, women who succeed in marrying such wealthy men would be well served reproductively, even more so than when polygyny is institutionalized, because polygyny results in each co-wife sharing her husband's bounty. The prediction therefore arises that when monogamy coincides with substantial differences among men with regard to their resources, women and their immediate families would compete among themselves to become a monogamously mated wife. Dowry should therefore occur in nonpolygynous but stratified societies rather than elsewhere. This prediction has been impressively confirmed: of the 994 societies that should not engage in dowry payments (because they are either polygynous or monogamous/polyandrous but not socioeconomically stratified), fewer than 1% do so. On the other hand, of 144 stratified, nonpolygynous societies, 54 (38%) employ dowries.[53] To conclude, dowry is roughly 50 times more common in societies that (1) are not polygynous and (2) are socioeconomically stratified than it is in other circumstances. And the reason, almost certainly, is that when there is a substantial wealth differential among men in monogamous societies, these men become analogous to women in polygynous societies: a comparatively scarce and limiting resource, worth competing for.

For an example at the other end of the socioeconomic scale, we have the Mukogodo people of Kenya. They have long occupied pretty much the bottom rungs of their geographic region with regard to wealth and prestige. As a result, their reproductive opportunities are comparably limited, even though there is considerable interbreeding between the Mukogodo and neighboring groups. The problem for the Mukogodo is that their neighbors regularly outcompete Mukogodo men for wives, having more goats, sheep, and cattle to offer as bride price. As a result, the Mukogodo place substantially less value on their sons—whose breeding options are limited—than on their daughters, who have a real prospect of marrying up. Mukogodo girls receive better medical care than their brothers, and their survival (at least to age five) is higher.[54] The sex ratio of live births favors girls as well, suggesting—but not proving—that in this case, infant boys may be the disproportionate victims of infanticide.

Cases such as these, in which wealthy, high-status families invest preferentially in sons and poorer, lower-status families, in girls, needn't reflect evolved traits per se. After all, although natural selection cannot see into the future, it can (and assuredly does) produce organisms who can do so. That's us.

Time now for some more direct attention to polygamy and parenting from the woman's perspective. To some extent, the polygyny threshold model has this virtue, although it may apply more to birds generally, in which females overfly male territories and—as far as we can tell—are free to choose among them.

Regrettably, when the question arises, "To be or not to be (one of a given man's co-wives)," female choice is less apparent among our own species.

Some observers—especially biologists, disposed to apply the polygyny threshold model to human beings—assume that polygynously mated women are roughly as happy as women who are monogamously married because their fitness is supposed to be equally maintained in polygynous or monogamous unions. A team of anthropologists, however, after looking through the Standard Cross-Cultural Sample, were able to identify 69 case studies of different social groups that spoke to the question of cooperation versus conflict among co-wives. Sexual and emotional conflicts—often involving children and the disposition of resources—were found to occur in more than 90% of the cases, whereas close friendships among the co-wives were reported in only 25%.[55]

Although as I have emphasized, human beings are perfectly good mammals, our species—like every species—has its peculiarities. In her book, *Mothers and Others*, anthropologist and primatologist Sarah Hrdy[56] has made a powerful case that people are not adapted for simple biparental care of their infants and children but rather our evolution shows the clear imprint of "alloparenting," assistance provided by aunts, uncles, cousins, siblings, grandparents, and so forth. In proportion as it takes—if not a village, then at least an assemblage of dedicated helpers—to rear a child, the existence of multiadult alloparenting would seem to free dominant and powerful men even more to be polygynists.

This leads me to the following apology, combined with a clarification. In proclaiming the wisdom of biparental care, I have implied that said parents are man and woman, presumably husband and wife. Biologically, this is justifiable because our reproduction indeed requires a genetic contribution from a man and a woman. Whatever induces the one more liable to defect from the reproductive partnership (in most cases, the man), to stay around instead and help—notably, confidence of parenthood—is likely to be a good evolutionary stratagem. At the same time, the more we learn about the crucial role of cooperative breeding, the more we see that "biparental" care can be provided by a range of parent-like figures, definitely including same-sex adult partners with commitment to each other.

There are two good reasons for pointing this out. One is simple adherence to scientific fact: unaided, single parenting is a hard slog, but certainly possible. Moreover, numerous arrangements exist whereby a single parent can get crucial help. The second reason is the political implications of hyping the supposed necessity of dual, husband–wife parenting. Thus, right-wing ideologues in the United States and elsewhere are prone to argue for government programs that "enhance marriage," "contribute to supporting the pair-bond," or whatever else it takes to sustain a long-term, heterosexual monogamous relationship as the best—ostensibly, in some cases, the only—way to rear healthy, happy, socially responsible children. In contrast, by recognizing the diversity of helpers found in human societies worldwide, a more expansive view of human alloparenting opens the door to accepting, encouraging, and otherwise supporting child care programs and systems that don't necessarily require traditional monogamous, heterosexual marriage, or indeed, marriage of any sort.

Sarah Hrdy points out that

> Contrary to the widely held dogma that only men who are certain of their
> paternity provide for young, in many widely separated corners of the world
> there exist customs and beliefs that help mothers elicit tolerance, protection,
> or assistance from men who are only possibly, rather than certainly, related.
> Among Eskimos, Montagnais-Naskapi, and some other North American
> Indian tribes, as well as among Central American people like the Siriono
> and many tribes in Amazonian South America well as across the ocean
> in parts of pre- and postcolonial west, central and east Africa, women are
> permitted or even encouraged to have sex with real or fictive brothers of their
> husbands. A range of innovations permits mothers in traditional societies
> from southwestern China and central Japan, as well as among people like the
> Lusi of Papua New Guinea and in areas of Polynesia to line up extra "fathers."
> Even in times and places renowned for patriarchal family structures, such as
> the Qing dynasty in China or in traditional India, desperately poor parents
> sometimes made ends meet by incorporating an extra man (preferably some
> kind of wage earner) into the marital unit.[57]

It appears that particularly when times are hard, women often enter into relationships that can be described as "sequentially polyandrous," thereby obtaining assistance from more than one man—even though only one will be the genetic father of
a given child. It can be a difficult and dicey process because sexual jealousy is rarely
far away. As described in chapter 1, a number of human societies maintain a belief in
"partible paternity," under which any man who has sexual intercourse with a woman
during her pregnancy is considered to be, at least in part, a father of the child who
eventually emerges. This, in turn, facilitates male acceptance of sequential polyandry.
As Hrdy recognizes

> Such mindsets are very different from those in Western society, where a long
> history of patrilineally transmitted resources leaves men preoccupied with
> genetic paternity and puts children whose paternity is in doubt at a serious
> disadvantage. But in partible-paternity societies, where relying on a single
> father is an even bigger than usual gamble, having several possible fathers has
> the opposite effect.

Among those human groups who practice (which is to say, believe in) partible
paternity are the Aché of Paraguay and the Bari of Venezuela.[58] (No one actually par-
takes of partible paternity; it is a biological impossibility.) Among both the Aché and
the Bari, children who have what is socially defined as many fathers are better off—or
at least, they experience lower mortality—than in societies whose traditions are more
genetically accurate but biologically as well as socially less rewarding. Partible paternity is a useful fiction. Just as belief in Santa Claus—although inaccurate—might
have the benevolent consequence of inducing children to be good, partible paternity,

along with the informal polyandry that it promotes, seems to have the benevolent consequence of getting men (even those who aren't "real" biological fathers) to be good fathers.

Hrdy concludes her discussion of partible paternity with a story both heartwarming and a useful corrective for those (such as, admittedly, myself) with a tendency to see the world through too strongly a biological lens:

> I am reminded of the long-ago Naskapi tribesman who was taken to task by an early Jesuit missionary in North America. Seeing the priest's dismay at the group's sexual promiscuity and uncertain paternity, the man responded: "Thou hast no sense. You French people love your own children; but we love all the children of our tribe." Spoken like a true cooperative breeder.

None of the foregoing should be taken to suggest that polygyny precludes polyandry.[59] Indeed, even if women are denied choice about joining the household of an already mated man, they still retain at least some options when it comes to who actually gets to father their children—which is to say, who they have sex with, at least on occasion (more on this in chapter 6). We have already considered the possibility that concealed ovulation evolved as a strategy that enables women to control their reproduction, or, at minimum to have some influence in this regard. Because the precise time of her ovulation is hidden, a woman would not be more closely guarded at that time than at any other, thereby giving her more leeway for an extracurricular rendezvous.

At the same time, the reality is that traditional harems, and polygynous households generally, are characterized by a great deal of sexual jealousy and what biologists call "mate guarding" on the part of the male and typically severe penalties for any of the wives—as well as their lovers—if they get caught.

Among mammals, as with birds, when males defend territories, these nearly always constitute terrain that is favored by females. Such cases of "resource defense polygyny"[60] have been especially well described for many species of ungulates, including waterbuck,[61] topi,[62] and pronghorn antelope.[63] Human beings, it must be emphasized, engage both in resource defense polygyny and in the alternative route: "female defense polygyny" of the sort exemplified by marmots[64] and gorillas.[65] In these cases, it is the females themselves who constitute the "territory" to be defended.

Sexual jealousy, especially male sexual jealousy, is so widespread that it is another of our human traits that is taken for granted and thus rarely interrogated. Almost certainly, it derives from internal fertilization and men's resulting inability to know for certain that their alleged children are "actually" (i.e., genetically) theirs.

Men have resorted to all sorts of social rules as well as physical interventions (chastity belts in the past, female circumcision even now in parts of the Islamic world) to allay the male fear of non-paternity. Of course, a mother and her immediate family have no such worries; they can be predicted, if anything, to concern themselves with reassuring the father and his family that he/they have not been cuckolded. An intriguing set of studies in Canada[66] and Mexico[67] found that relatives of the mother are especially likely to emphasize the extent to which a newborn baby resembles the father.

With polyandry at issue, let's take a look at one of the very few human societies in which full-scale "classical polyandry" is institutionalized. Sadly, it turns out that even here, female choice is subordinated to male interests. The Tre-ba people of Tibet and northern Nepal are relatively prosperous but land-poor peasants for whom polyandry seems to be a mechanism that prevents their limited arable land from being subdivided, via inheritance, into plots so small as to be economically useless. (In Europe, by contrast, landholdings were kept of adequate size largely via primogeniture.)

There is only one marriage in each generation of Tre-ba families. Brothers get together and *obtain* one wife, who they share sexually, thereby avoiding partition of their family landholdings. But even here, women aren't socially dominant. The shared wife is expected to do much of the housework and essentially all the child care. And it is the men who get together to purchase a wife (via bride price) rather than a wife choosing to affiliate with a particular set of multiple husbands. Significantly, when a family's offspring consists of daughters and no sons, even the Tre-ba become polygynists, with the daughters (sisters) taking one husband. Either way, the land remains undivided, and women's choice is circumscribed.[68]

The fact that among the Tre-ba, classical polyandry is fraternal is further evidence of the social impact of men's biological interest. Thus, even though a Tre-ba co-husband isn't assured of fathering the household's children, he (like those Tasmanian native hens we met earlier) is at least guaranteed to be an uncle. Fraternal polyandry is essentially the mirror-image of sororal polygyny in that when siblings share a husband or wife, there is predictably less same-sex conflict than if the co-spouses are unrelated and therefore have no genetic interest in each other's offspring.

Interestingly, among those social groups that—unlike the Tre-ba—engage in "informal polyandry," senior husbands typically attempt to restrict the sexual access of junior husbands, whose mounting dissatisfaction often induces them to leave when new marital prospects materialize. It isn't clear to what extent comparable sexual competition among Tre-ba co-husbands generates conflict and instability. On one hand, some such difficulties can be predicted, although on the other, we can also predict that troubles of this sort would likely be leavened by the fact that the men in question are full siblings.

The role of genetic interest seems to have a powerful effect on another one of those human traits that is taken for granted, but which—like concealed ovulation—is biologically unusual and repays greater attention. The issue here is menopause, the pan-human universal that a woman's reproductive capacity terminates when she approaches age 50. Given the evolutionary payoff to producing just one more offspring, this closing of the ovulatory faucet is a genuine biological mystery, all the more so because postmenopausal women typically have several more decades of healthy life ahead of them. Moreover, it is extremely unusual among animals: in nearly every species, females keep breeding pretty much until they are on their last legs.

Elsewhere I have reviewed the various hypotheses that have been proposed to explain this evolutionary enigma.[69] Here, I will focus on just one, which seems not only the most viable, but is closely associated with human polygamy. Begin by stipulating that as the human body ages, it is less capable of withstanding the rigors of pregnancy, childbirth, and lactation. Even here, simple arithmetic shows that selection would favor last-ditch reproduction, even by elderly women, and even if morbidity and mortality eventually intervened. There is, however, an important exception: insofar as aging women have an opportunity to improve their own inclusive fitness by contributing positively to the reproductive success of their children by assisting in the care of their children's children—the menopausal woman's grandchildren.

Hence, the "grandmother hypothesis,"[70] which suggests that women are selected to stop reproducing when they can do more to project their genes into the future by helping their grandchildren than if they were to risk attempting to make more children of their own. Note that women undergo menopause at the age when their own children are beginning to reproduce. Moreover, there is a growing wealth of information showing that—especially in traditional societies—grandmothers not only work hard on behalf of their grandchildren (sometimes providing more direct calories than do the mothers), but that young children in households with contributing grandmothers do significantly better than when grandmothers are absent.[71]

The grandmother effect could operate in monogamous households as well as in polygynous ones. But certainly it isn't excluded from the latter. Moreover, the possibility arises that for all the benevolence of the grandmother hypothesis, situations of polygyny provide an added and more pernicious opportunity for a grandmother's involvement: directing resources asymmetrically toward her own grandchildren and even, perhaps, injuring those other children to whom she is not related.

Menopause is limited to women. Although men's reproductive competence declines somewhat with age, there is no "male menopause" comparable to women's. We've already noted that modern Western societies that are formally monogamous turn out to be effectively polygamous, especially when considering remarriages and the formation of second and third families. There is a substantial gender asymmetry: men remarry more than do women and have more children in their subsequent marriages.[72] This is due to the fact that by the time of a second or third marriage, the participants are older and fertility declines dramatically with age among women but not among men.

The consequences haven't been altogether benevolent. It has long been known that older parents are more liable to have children with genetic birth defects. The great majority of these are due to older fathers. The reason is straightforward biology: a girl is born with all the eggs she will ever produce; thus, they are all equally old, or young. By contrast, sperm are made afresh throughout a man's life, from sperm-generating tissue that continue to reproduce as a man ages. As a result, the older the father, the more opportunities arise for errors (mutations) to intervene. "A thirty-six year-old father passes on twice as many [mutations] as a twenty-year-old; a seventy year-old, eight times as many."[73] If not for polygyny and its resulting tendency for old men to reproduce with young women, our species would have substantially fewer birth defects.

H. L. Mencken, the "sage of Baltimore," provided his own inimitable explanation of why men are more likely than women to marry late in life. If you haven't encountered Mencken's delightful curmudgeonly writing, which comes from the days when marriage was supposed to be something for which women yearned and men sought to avoid, you are in for a treat (some advice, however: before proceeding, check any feminist outrage as well as your evolutionary biology—indeed, any biology whatsoever—at the door):

The marriage of a first-rate man, when it takes place at all, commonly takes place relatively late. He may succumb in the end, but he is almost always able to postpone the disaster a good deal longer than the average poor clodpate, or normal man. If he actually marries early, it is nearly always proof that some intolerable external pressure has been applied to him . . . The reasons which impel him to yield even then are somewhat obscure, but two or three of them, perhaps, may be vaguely discerned. One lies in the fact that every man, whether of the first-class or of any other class, tends to decline in mental agility as he grows older, though in the actual range and profundity of his intelligence he may keep on improving until he collapses into senility. Obviously, it is mere agility of mind, and not profundity, that is of most value and effect in so tricky and deceptive a combat as the duel of sex. The aging man, with his agility gradually withering, is thus confronted by women in whom it still luxuriates as a function of their relative youth. Not only do women of his own age aspire to ensnare him, but also women of all ages back to adolescence. Hence his average or typical opponent tends to be progressively younger and younger than he is, and in the end the mere advantage of her youth may be sufficient to tip over his tottering defences. This, I take it, is why oldish men are so often intrigued by girls in their teens. It is not that age calls maudlinly to youth, as the poets would have it; it is that age is no match for youth, especially when age is male and youth is female.

Another reason for the relatively late marriages of superior men is found, perhaps, in the fact that, as a man grows older, the disabilities he suffers by marriage tend to diminish and the advantages to increase. At thirty a man is terrified by the inhibitions of monogamy and has little taste for the so-called comforts of a home; at sixty he is beyond amorous adventure and is in need of creature ease and security. What he is oftenest conscious of, in these later years, is his physical decay; he sees himself as in imminent danger of falling into neglect and helplessness. He is thus confronted by a choice between getting a wife or hiring a nurse, and he commonly chooses the wife as the less expensive and exacting. The nurse, indeed, would probably try to marry him anyhow; if he employs her in place of a wife he commonly ends by finding himself married and minus a nurse, to his confusion and discomfiture, and to the far greater discomfiture of his heirs and assigns. . . . If men, at the normal mating age, had half as much to gain by marriage as women gain, then, all men would be as ardently in favour of it as women are.[74]

The biomedical reality that men can continue to reproduce into old age probably has been both an outcome as well as a driver of polygyny. Part of the natural progression of polygyny in overtly harem-forming societies is for some men to accumulate wives as, with age, they accumulate power, prestige, and resources. It is even possible that a major reason why men don't undergo menopause is precisely because such accumulation regularly occurred in human prehistory, so that some men (successful polygynists) had the opportunity to reproduce into old age, whereupon those who didn't experience male menopause were more fit than were those that did.

Finally, a bit about polygyny and childrearing generally. Bobbi Low, of the University of Michigan, tested evolutionary predictions connecting polygyny and childrearing practices across different human groups. Among these predictions were:

1. Sons and daughters ought[75] to be trained differently in all societies.
2. Sons ought to be inculcated more strongly than daughters in competitive behaviors that are likely to result in the acquisition and control of resources useful in getting mates.
3. The more polygynous the society, the more boys ought to be taught to strive.
4. In stratified societies, whether the stratification is by wealth or heredity, the impact of striving on reproductive success for men may be muted. That is, whereas women may marry up to a higher class, men are seldom able to do so. Furthermore, stratified societies frequently have rules that limit not only the males' status but the number of wives allowed to men of different statuses. Thus, the more stratified the society (the more reduced the correlation between striving and possible reproductive payoffs to males), the less boys are taught to strive openly (though sneaky competition may exist).
5. Stratified polygynous societies with hypergyny ought to show stronger inculcation of daughters in sexual restraint and obedience than in other societies, because these increase a woman's apparent value to prospective high-status husbands.[76]

All of these predictions were supported by the available cross-cultural data. Professor Low concluded that male training is significantly skewed, relative to female training, with regard to aggression, dutifulness, fortitude, obedience, and toughness. This finding—although totally consistent with what we know from biology—is nonetheless troublesome for anyone sensitive to the regrettable human tendency for patriarchy and sexism, reflected, for example, in the now appropriately unfashionable nursery rhyme:

What are little boys made of?
Snakes and snails
And puppy-dogs' tails,
That's what little boys are made of.
What are little girls made of?
Sugar and spice
And everything nice,
That's what little girls are made of.

Here and elsewhere we are faced with a chicken-and-egg dilemma: our polygynous background is both cause and effect of many of our deeply human traits. For example, male–female differences in confidence of relatedness to offspring, which in turn selects for female lactation, contributes to polygyny by freeing men to compete for wives rather than focusing on their children.[77] At the same time, however, polygyny is a result (an effect) of this same fact: that men, because of their mammalian biology, are "liberated" from mandatory child care.

Zoologist Tim Clutton-Brock included the following acknowledgment in his book, *The Evolution of Parental Care*: "My greatest debt is to my wife, . . . [who] looked after our children while I wrote about parental care."[78]

Notes

1. Note that there is no implication here that such confidence need be conscious. Any genetically influenced tendency among males to invest in someone else's offspring would be selected against, relative to tendencies to concentrate paternal investment in one's own progeny.
2. Low, B. (1992). Men, women, resources and politics. In J. M. G. van der Denned (Ed.), *The Nature of the sexes*. Groningen, Holland: Origin Press.
3. Daly, M., & Wilson, M. (1978). *Sex, evolution, and behavior*. North Scituate, MA: Duxbury Press.
4. Ralls, K. (1976). Mammals in which females are larger than males. *Quarterly Review of Biology, 51*, 181–225.
5. Marlowe, F. (2000). Paternal investment and the human mating system. *Behavioral Processes, 51*, 45–61.
6. Ehrenreich, B. (1987). *The hearts of men: American dreams and the flight from commitment*. New York: Doubleday Anchor.
7. Sear, R., & Mace, R. (2008). Who keeps children alive? A review of the effects of kin on child survival. *Evolution and human behavior, 29*, 1–18.
8. Orians, G. H. (1969). On the evolution of mating systems in birds and mammals. *American Naturalist, 103*, 589–603.
9. Barash, D. P. (1976). Some evolutionary aspects of parental behavior in animals and man. *American Journal of Psychology, 89*, 195–217.

10. In this case, the word "selected" does double duty, referring both to an individual female's choice and to the action of natural selection.

11. Alatalo, R. V., Lundberg, A., & Glynn, C. (1986). Female pied flycatchers choose territory quality and not male characteristics. *Nature, 323*, 152–153.

12. Lenington, S. (1980). Female choice and polygyny in red-winged blackbirds. *Animal Behaviour, 28*, 347–351.

13. Pleszczynska, W. K. (1978). Microgeographic prediction of polygyny in the lark bunting. *Science, 201*, 935–937.

14. Travis, S. E., Slobodchikoff, C. N., & Keim, P. (1995). Ecological and demographic effects on intraspecific variation in the social system of prairie dogs. *Ecology, 76*, 1794–1803.

15. Downhower, J., & Armitage, K. B. (1971). The yellow-bellied marmot and the evolution of polygyny. *American Naturalist, 105*, 355–370.

16. Irons, W. (1979). Cultural and biological success. In N. A. Chagnon & W. Irons (Eds.), *Natural selection and social behavior*. North Scituate, MA: Duxbury.

17. Flinn, M. V. (1986). Correlates of reproductive success in a Caribbean village. *Human Ecology, 14*, 225–243.

18. Turke, P. W., & Betzig, L. (1985). Those who can do: Wealth, status, and reproductive success on Ifaluk. *Ethology and Sociobiology, 6*, 79–87.

19. Hill, J. (1984). Prestige and reproductive success in man. *Ethology and Sociobiology, 5*, 77–95.

20. Kempenaers, B., Everding, S., Bishop, C., Boag, P., & Robertson, R. J. (2001). Extra-pair paternity and the reproductive role of male floaters in the tree swallow (*Tachycineta bicolor*). *Behavioral Ecology and Sociobiology, 49*, 251–259.

21. Lee, R. (1979). Politics, sexual and nonsexual, in an egalitarian society. In E. Leacock & R. Lee (Eds.), *Politics and history in band societies*. Cambridge, England: Cambridge University Press.

22. Quoted in Hrdy, S. (1999). *Mother nature: Maternal instincts and how they shape the human species*. New York: Ballantine Books.

23. For example, Strassman, B. (1997). Polygyny as a risk factor for child mortality among the Dogon. *Current Anthropology, 38*, 688–695.

24. Variations on this procedure (widely considered abhorrent as well as medically dangerous) are performed in more than two dozen countries in Africa and the Middle East.

25. Hrdy, *Mother Nature*.

26. Isaac, B. L. (1980). Female fertility and marital form among the Mende of rural upper Bambara chiefdom, Sierra Leone. *Ethnology, 19*, 297–313.

27. Strassman, B. (2003). Social monogamy in a human society: Marriage and reproductive success among the Dogon. In B. Reichard & C. Boesch (Eds.), *Monogamy: Mating strategies and partnerships in birds, humans and other mammals*. Cambridge, England: Cambridge University Press.

28. Flinn, M. V., & England, B. (1995). Childhood stress and family environment. *Current Anthropology, 36*, 854–866.

29. Dorjahn, V. R. (1958). Fertility, polygyny and their interrelations in Temne society. *American Anthropologist, 60*, 838–860.

30. Borgerhoff Mulder, M. (1990). Kipsigis women's preference for wealthy men: Evidence for female choice in mammals? *Behavioral Ecology and Sociobiology, 17*, 255–264.

31. Bove, R., & Valeggia, C. (2009). Polygyny and women's health in sub-Saharan Africa. *Social Science and Medicine, 68*, 21–29.

32. Jankowiak, W., Sudakov, M., & Wilreker, B. C. (2005). Co-wife conflict and cooperation. *Ethnology, 44,* 81–98.

33. Ridpath, M. G. (1972). The Tasmanian native hen, *Tribonyx mortierii. CSIRO Wildlife Research, 17,* 1–118.

34. Chisholm, J. S., & Burbank, V. (1991). Monogamy and polygyny in southeast Arnhem Land: Male coercion and female choice. *Ethology and Sociobiology, 12,* 291–313.

35. Egboh, E. O. (1972). Polygamy in Iboland. *Civilizations, 22,* 431–444.

36. Eagly, A. H., & Wood, W. (1999). The origins of sex differences in human behavior: Evolved dispositions or social roles? *American Psychologist, 54,* 408–423.

37. Schmitt, D. P., Youn, G., Bond, B., Brooks, S., Frye, H., Johnson, S., . . . & Stoka, C. (2009). When will I feel love? The effects of culture, personality, and gender on the psychological tendency to love. *Journal of Research in Personality, 43,* 830–846; Lippa, R. A. (2009). Sex differences in sex drive, sociosexuality, and height across 53 nations: Testing evolutionary and social structural theories. *Archives of Sexual Behavior, 38,* 631–651.

38. Tadinac, M., & Hromatko, I. (2007). Own mate value and relative importance of a potential mate's qualities. *Studia Psychologica, 49,* 251–264; Buss, D. M., & Shackelford, T. K. (2008). Attractive women want it all: Good genes, economic investment, parenting proclivities, and emotional commitment. *Evolutionary Psychology, 6,* 134–146.

39. For instance, if in our example the sex ratio changes so that there are more than 50% males, then the average payoff to a male decreases by precisely the degree of the imbalance; this, in turn, selects for producing more females. How many? Just enough to equalize the sex ratio.

40. Trivers, R. L., & Willard, D. E. (1973). Natural selection of parental ability to vary the sex ratio of offspring. *Science, 179*(4068), 90–92.

41. Proffitt, K. M., Garrott, R. A., & Rotella, J. J. (2008). Variation in offspring sex ratio among individual Weddell seal (*Leptonychotes weddellii*) females of different quality. *Behavioral Ecology and Sociobiology, 62,* 1679–1687.

42. Holand, Ø., Mysterud, A., Røed, K. H., Coulson, T., Gjøstein, H., Weladji, R. B., & Nieminen, M. (2006). Adaptive adjustment of offspring sex ratio and maternal reproductive effort in an iteroparous mammal. *Proceedings of the Royal Society B: Biological Sciences, 273*(1584), 293–299.

43. Roche, J. R., Lee, J. M., & Berry, D. P. (2006). Pre-conception energy balance and secondary sex ratio—Partial support for the Trivers-Willard hypothesis in dairy cows. *Journal of Dairy Science, 89,* 2119–2125.

44. Brown, G. R. (2001). Sex-biased investment in nonhuman primates: Can Trivers & Willard's theory be tested? *Animal Behaviour, 61,* 683–694.

45. Dickemann, M. (1979). Female infanticide, reproductive strategies and social stratification: A preliminary model. In N. Chagnon & W. Irons (Eds.), *Evolutionary biology and human social behavior.* North Scituate, MA: Duxbury Press.

46. Autor, D., Figlio, D., Karbownik, K., Roth, J., & Wasserman, M. (2015). Family disadvantage and the gender gap in behavioral and educational outcomes. http://www.ipr.northwestern.edu/publications/papers/2015/ipr-wp-15-16.html.

47. Kanazawa, S. (2005). Big and tall parents have more sons: Further generalizations of the Trivers–Willard hypothesis. *Journal of Theoretical Biology, 235,* 583–590.

48. Murdoch, G. P. (1976). *Ethnographic Atlas.* Pittsburg, PA: University of Pittsburg Press.

49. Murdoch, G. P., & White, D. (1969). Standard cross-cultural sample. *Ethnology, 8,* 329–369.

50. Although the SCCS is smaller, it represents societies that are—insofar as possible— geographically and linguistically quite separate from each other and therefore more likely to be composed of what researchers call truly independent "data sets."

51. Borgerhoff Mulder, M., & Turke, P. (1988). Kipsigis bridewealth payments. In L. Betzig, M. Borgerhoff Mulder, & P. Turke (Eds.), *Human reproductive behaviour: A Darwinian perspective.* Cambridge, England: Cambridge University Press.

52. Gaulin, S. J. C., & Boster, J. S. (1990). Dowry as female competition. *American Anthropologist, 92,* 994–1005.

53. Ibid.

54. Cronk, L. (1989). Low socioeconomic status and female-biased parental investment: The Mukogodo example. *American Anthropologist, 91,* 414–429.

55. Jankowiak, W., Sudakov, M., & Wilreker, B. C. (2005). Co-wife conflict and co-operation. *Ethnology, 44,* 81–98.

56. Hrdy, S. B. (2009). *Mothers and others: The evolutionary origins of mutual understanding.* Cambridge, MA: Harvard University Press.

57. Ibid.

58. Hill, K., & Hurtado, A. M. (1996). *The ecology and demography of a foraging people.* New York: Aldyne de Gruyer; Beckerman, S., & Valentine, P. (2002). *The theory and practice of partible paternity in South America.* Gainesville: University Press of Florida.

59. Or *vice versa.*

60. Emlen, S., & Oring, L. (1977). Ecology, sexual selection, and the evolution of mating systems. *Science, 197,* 215–223.

61. Spinage, C. A. (1969). Territoriality and social organization of the Uganda defassa waterbuck, *Kobus defassa ugandae. Journal of Zoology, 159,* 329–361.

62. Gosling, L. M. (1991). The alternative mating strategies of male topi, *Damaliscus lunatus. Applied Animal Behaviour Science, 29,* 107–119.

63. Byers, J. A. (1997). *American pronghorn: Social adaptations and the ghosts of predators past.* Chicago: University of Chicago Press.

64. Barash, D. P. (1989). *Marmots: Social behavior and ecology.* Stanford, CA: Stanford University Press.

65. Dixson, A. F. (2009). *Sexual selection and the origins of human mating systems.* Oxford, England: Oxford University Press.

66. Daly, M. & Wilson, M. (1982). Whom are newborn babies said to resemble? *Ethology and Sociobiology, 3,* 69–78.

67. Regalski, J. M., & Gaulin, S. J. C. (1993). Whom are Mexican infants said to resemble? Monitoring and fostering paternal confidence in the Yucatan. *Ethology and Sociobiology, 14,* 97–113.

68. Goldstein, M. C. (1971). Stratification, polyandry, and family structure in central Tibet. *Southwestern Journal of Anthropology, 27,* 64–74.

69. Barash, D. P., & Lipton, J. E. (2009). *How women got their curves, and other just-so stories.* New York: Columbia University Press; Barash, D. P. (2012). *Homo mysterious: Evolutionary puzzles of human nature.* New York: Oxford University Press.

70. Voland, E., Chasiotis, A., & Schiefenhövel, W. (Eds.). (2005). *Grandmotherhood: The evolutionary significance of the second half of female life.* New Brunswick, NJ: Rutgers University Press.

71. Hawkes, K. (2004). Human longevity: The grandmother effect. *Nature, 428*(6979): 128–129.

72. Essock-Vitale, S., & McGuire, M. T. (1988). What 70 million years hath wrought: Sexual histories and reproductive success of a random sample of American women. In L. Betzig, M. Borgerhoff-Mulder, & P. Turke (Eds.), *Human reproductive behavior: A Darwinian perspective*. Cambridge, England: Cambridge University Press.
73. Konner, M. (2015). *Women after all*. New York: W. W. Norton.
74. Mencken, H. L. (1927). *In Defense of Women*. New York: A. A. Knopf.
75. Note that this is not an ethical "ought" but rather simply a way of stating a deduced prediction.
76. Low, B. (1989). Cross-cultural patterns in the training of children: An evolutionary perspective. *Journal of Comparative Psychology, 103*, 311–319.
77. An alternative and equally valid perspective is that because men lack women's confidence in genetic relatedness to their offspring, men are kicked out of the Eden of nursing their infants; and so, making the best of a bad situation, they find themselves seeking additional mating opportunities.
78. Cllutton-Brock, T. H. (1991). *The evolution of parental care*. Princeton, NJ: Princeton University Press.

5

What About Monogamy?

In Nora Ephron's autobiographical movie, *Heartburn*, Meryl Streep's character complains to her father about her husband's many infidelities. His response: "You want monogamy? Marry a swan!" Now we know that even swans aren't monogamous. (At least, not rigorously so.)

Given that strict monogamy is biologically unnatural for people, too, this question arises: why is it the preferred—even legally and socially mandated—marital lifestyle in so much of the modern world? This is an open question. In this chapter I will suggest some possible, speculative answers, but no more than that.

Monogamy isn't a yes–no, black-or-white thing. Neither is polygyny or polyandry. Instead, these mating systems occupy a sliding scale, depending on the details of sexual behavior. When studying animals, we biologists can intrude, eavesdrop, and otherwise stick our noses into their (literal) affairs without worrying about the ethics of invading their privacy. Not so with people, whose intimate lives can only be interrogated by surveys. And there is nothing people lie about more than their sex lives; the ironic result is that we know more about the sexual shenanigans of chimpanzees, bonobos, or song sparrows than about *Homo sapiens*.

Even then, we don't know as much as one might hope. Our ignorance is especially great when it comes to mammals (compared with birds, for example), because the majority of mammals are not only shy but nocturnal: most mammals are either bats or rodents. Nonetheless, a detailed analysis of the evolutionary relationships among the existing mammal species about which basic natural history information is available concluded that monogamy arose independently 61 times, giving rise to 229 different species[1]—which might seem like a lot, but comprises fewer than 9% of all mammals. As mammals go, primates are unusually predisposed to monogamy, but this still constitutes only 29% of species, and even then, with lots of sexual dalliance occurring as well. Gibbons, for example, are socially monogamous and were assumed to be sexually monogamous as well until detailed field studies found that adult females periodically mated with neighboring males, who may or may not themselves be "monogamously" mated.[2]

The common pattern is that among those mammals that are at least somewhat inclined toward monogamy, females are comparatively solitary and widely dispersed, mostly because their food sources are spread out. In short, the research concluded that when female mammals live pretty much by themselves, because there isn't enough food in any one place to support a "sisterhood," males typically end up with only one mate—most of the time—simply because even the most ardent male can only be in one place at a time.[3] Ditto for females.

By contrast, another recent study that looked specifically at ostensibly monogamous nonhuman primates concluded differently: their most common shared trait appeared to be a risk of infanticide.[4] Under these circumstances, it is possible that males, as well as females, would have been more fit if they stayed with the mother of their offspring, so as to defend the latter. It isn't clear how, if at all, human beings fit into these competing scenarios, which have become quite controversial among the biologists involved. Nonetheless, it is undeniable that our own species is unusual in that small breeding units—families, whether monogamous or polygamous—exist within larger social groupings. Among nonhuman primates, those that are or that approximate monogamy—such as most species of gibbons and marmosets—live apart from other such social units.

Aside from the messy, misty question of our evolutionary heritage (technically, the "phylogenetic history" of human mating systems), any serious assessment of *Homo sapiens* must confront a definitional problem, something that goes beyond lexicographical fussiness. Consider a human marriage that is ostensibly monogamous but in which the husband has a single, one-night stand: or two, or a hundred, but the pair nonetheless remains together. Ditto for the woman. At what point is their union no longer monogamous?

To be biologically meaningful, we should ideally assess the degree of polygamy by calculating the ratio of male to female variance in reproductive success. As already noted, a ratio of one would indicate genuine monogamy, at least in a way that connects with evolutionary processes; a ratio of less than one defines polyandry; and greater than one, polygyny. But what about variance in number of sexual partners? (Given the availability of birth control, there is less and less correlation between actual copulation frequency and number of offspring produced; the former speaks to sexual behavior, whereas the latter speaks to its evolutionary significance.)

Perhaps it's a fool's errand to struggle too much with something as squishy as human mating systems, and to concern ourselves excessively with whether a given relationship is "genuinely monogamous." We might be well advised to worry less about trying to define monogamy and concern ourselves more with trying to understand it, since at some level, we all know what the word means.

Any consideration of monogamy—especially one looking for measurable evidence—must confront the distressing reality that if anything, it is likely to be substantially less frequent than reported. For one thing, there is the problem of how to define monogamy, and the fact that in its idealized, extreme state—just one sexual partner over one's entire life—it is exceedingly rare. How rare? Impossible to know for sure. As already mentioned, there is nothing about which people are less honest and forthcoming than their sex lives. For another, serial monogamy—the most common identified human mating system, at least in the industrialized West—could just as well be labeled "serial polygyny," in which case monogamy, we must conclude, is rarer yet. Recall the movie *March of the Penguins*, a heartwarming account of seemingly monogamous fidelity among emperor penguins. The reality, however, is that marital commitment among these admirable creatures lasts only for one breeding season. They are serially monogamous—or serially polygamous, take your pick: next year, each partner will establish a new pair bond with someone else.

Moreover, it turns out that even among human societies in which monogamy is officially mandated, reality often departs from sexual monogamy. Social monogamy is one thing; sexual monogamy is another. Take, for example, the case of ancient Rome, which has been analyzed in revealing detail by evolutionary anthropologist/ biologist/historian Laura Betzig.[5] Her work is an excellent example of what social anthropologist Clifford Geertz[6] called "thick analysis," involving a rich combination of historical, anecdotal, literary, and culturally informed information whenever clear-cut, quantitative data are unavailable. There is no doubt that Roman society was determinedly monogamous: at least, even Roman aristocrats and plutocrats took only one legal wife—at a time.[7] Betzig's work makes it clear, however, that despite the formalization of monogamy, upper-class Romans were flagrantly polygynous, and presumably polyandrous as well.

This is dispiriting perhaps, but shouldn't be surprising. The insights of evolutionary biology plus all available cross-cultural evidence shows that power and polygyny (either de jure or de facto) have always gone together. As Betzig summarizes

> The evidence across cultures is consistent. In the simplest societies, like the !Kung in Botswana or the Yanomamo in Venezuela, the strongest men typically kept up to ten women; in medium-sized societies that organized above the local level, like the Samoans and other Polynesians, men at the top kept up to a hundred women; and in the biggest societies, including the "pristine" empires in Mesopotamia and Egypt, India and China, Aztec Mexico and Inca Peru, and in many empires that came later, powerful men kept hundreds, or thousands, or even tens of thousands of women—along with one, or two, or three at most, legitimate *wives*; lesser men kept progressively fewer women.

Betzig provides abundant, rich, and thick descriptions of the sexual indulgences of the (ostensibly monogamous) Roman emperors Julius Caesar, Augustus, Tiberius, Caligula, even Claudius (the supposedly benign and introverted), and—no slouch in the multiple matings department—Nero, followed by a slew of lesser known but only somewhat less lecherous emperors. Marcus Aurelius, the Stoic philosopher emperor, and Vespasian, described by the historian Suetonius as "old and feeble," may have been the only exceptions. Particularly worth repeating from Betzig's trove of polygynous Romaniana is this verse, created and sung by Julius Caesar's soldiers, about their emperor and military leader: "Home we bring our bald whoremonger;/ Romans, lock your wives away!/All the bags of gold you lent him/Went his Gallic tarts to pay." Also this rather understated account of one of Rome's lesser known emperors, Elagabalus: he "never had intercourse with the same woman twice, except with his wife."

Further evidence for Roman polygynous mating comes from several additional sources, such as the immense literature on the keeping of slaves, who were often numbered in the hundreds and sometimes thousands. Slave women didn't merely do the laundry; they were frequently praised for their childbearing ability, as well as their physical attractiveness. Moreover, it is highly unlikely that they were valued only for their economic usefulness, namely their capacity to mate with male slaves and thus breed additional slaves. Much like slave owners in the antebellum American South, it is not just probable but certain that masters often copulated with their female slaves. Indeed, it is very likely that their sexual and reproductive services were much of what these enslaved people were *for*. Sexual access on the part of commoners to enslaved women was often as restricted as access to traditional, Oriental-style harems, which would hardly be expected if a primary function of these women was merely to reproduce, never mind with whom. As Betzig concludes, "Despotism explains differential reproduction. Power predicts polygyny, in Rome, and apparently everywhere else."

Which brings us back to the question, why monogamy?

One cogent possibility revolves around the fact that human infants are born in such a profoundly needy and immature state, requiring so much subsequent parental investment that the committed involvement of two parents conveys a considerable fitness benefit. By contrast, in polygynous species, there is relatively little paternal involvement because the harem-holding male must divide his attention across his numerous mates and their offspring—not to mention spending much of his time and energy repelling the takeover or sneak-in efforts of other males.

In addition, contrary to what most people assume—and despite its drawbacks for co-wives, as described in the last chapter—polygyny is actually a better deal for women than for most men. Thus, under polygyny, all women get mates (at least in theory) compared with only a very small proportion of men; monogamy is therefore an equalizing and democratizing system, for men, in that it immensely increases the likelihood that they, too, will be able to marry.

My experience has been that when men learn about polygyny, a frequent response is to libidinously lick their chops and imagine how delightful to be part of such a mating system. They assume, of course, that they would be a harem master. The problem is analogous to those who claim to remember their past lives: nearly always, such people proclaim themselves to have been the likes of Napoleon, Julius Caesar, Cleopatra, or perhaps Marie Curie, whereas they would most likely have been a low-level recruit in the armies of the first two, or one of Cleopatra's slaves, or at best, a maid who swept the floor of Madame Curie's lab. In a polygynous system, most men weren't polygynists themselves; that august position was reserved for a very small minority.

Accordingly, dear reader, it probably wouldn't have been you (or me), although hope springs eternal. It has been suggested that the allure of polygyny has been one of the factors generating resistance, especially in Africa, to Christian missionary efforts.[8]

There is a long tradition, at least in the West, of decrying monogamous marriage, especially from the perspective of husbands whose sexual freedom is ostensibly diminished by marriage, even as responsibilities are multiplied. "In our monogamous part of the world," wrote the famously gloomy philosopher Arthur Schopenhauer, "to marry means to halve one's rights and double one's duties." He was a lifelong bachelor.

Nonetheless, it is unmarried men—far more numerous than the harem-holding polygynists—who are the big losers in any harem system. Earlier, I noted George Bernard Shaw's recognition that polygyny could serve the interest of women by associating them with wealthier and more powerful ("first-rate") men rather than consigning them to "third-rate" ones. At the same time, Shaw recognized that polygyny was a recipe for discontent—especially on the part of men. Polygyny, he wrote "when tried under modern democratic conditions . . . is wrecked by the revolt of the mass of inferior men who are condemned to celibacy by it." Monogamy may therefore have emerged as a sop to men, reducing the number consigned to frustrated bachelorhood, in a kind of unspoken social bargain whereby powerful men gave up the overt perquisites of polygyny in return for obtaining a degree of social peace and harmony. Maybe it's not religion that is the opiate of the masses, as Marx claimed, but monogamy.

On the other hand, it is also possible (although as we shall see, unlikely) that monogamy is one of those social constructs that do more harm than good—like female circumcision or, in the United States, the Electoral College system. In *Civilization and Its Discontents*, Freud pointed out that civilization is founded on the suppression of the instincts, by which he particularly meant the "id," a combination of sexual voraciousness and personal violence. *Civilization and Its Discontents* was Freud's most pessimistic book because in it, he recognized that human beings could not function without civilization, and yet civilized life required that the "id" be denied, resulting in neuroses: damned if we do, damned if we don't.

Although Freud wasn't referring specifically to polygamy, he could have been. Or at least, an updating of Freud's thought in light of material in this book might be titled *Monogamy and Its Discontents* because a civilization that insists on monogamy requires that both the male predisposition for polygyny and the female penchant for polyandry are inhibited—or alternatively, lied about. It may nonetheless have arisen and persisted because the alternative—*Polygyny and Its Discontents*—is too socially disruptive.

So far, I've hinted at two major theories that seek to account for the prominence of monogamy in modern times: increasing the number of committed parent figures and achieving a degree of social equality. In the rest of this chapter, I will explore these two general hypotheses, along with a third: maybe monogamy is due to love. Seriously.

L et's start with parenting. Although monogamy isn't altogether "natural" for human beings, it is definitely a system that works for our species, at least in part because it meets some of our most crucial needs remarkably well. A cynical view would point out that *Homo sapiens* is, after all, a notably adaptive organism, able to adjust to all sorts of ecological situations (literally from the arctic to the tropics, from sea level to mountains, and pretty much everywhere in between), as well as a wide range of social arrangements. So maybe we accept monogamy just like we accept the need to propitiate a bossy in-law, or live with a noisy neighbor. The likelihood, however, is that monogamy is not just a social imposition that human beings find barely tolerable but something that fits many of our needs just fine.

A major reason for this fit is that because of our unique parenting needs, monogamy literally contributes to evolutionary fitness. As already noted, human infants are helpless at birth, requiring an enormous amount of adult investment, not simply to protect and nourish them but to see that they obtain the training, experience, and education required for success. And even though single parenting is definitely feasible, the data are overwhelming that cross-culturally, young children do better in proportion as more adults are concerned with their welfare. It may or may not take a village to raise a child, but committed adults—of whatever gender—are a definite plus.[9]

How then to obtain such a commitment? There are very few deadbeat moms because a woman who has given birth is definitely genetically connected to her child, as well as hormonally and emotionally predisposed to transition to motherhood. For every mother there is a father, but—for reasons we have already discussed, namely, the genetic uncertainty that comes with internal fertilization—fatherhood is a more iffy proposition. A special charm of monogamy is that by enhancing a man's confidence in paternity, his predisposition to act paternally is enhanced in turn.

It doesn't require a trained biologist to recognize the benefits of two committed adults when it comes to rearing children. As long ago as 1762, Jean-Jacques Rousseau wrote in his treatise, *Emile*, that

> Although a man does not brood like a pigeon, and although he has no milk to suckle the young, . . . his children are feeble and helpless for so long a time, that mother and children could ill dispense with the father's affection and the care which results from it.

This assertion is notably hypocritical, given that Rousseau himself fathered five children by his mistress, Thérèse Le Vasseur, insisting that each one be abandoned to foundling homes! Nevertheless, it is biologically and socially accurate. Pigeons are interesting in this regard because birds generally are much more inclined to monogamy (at least social monogamy) than are mammals. And this, in turn, makes sense in terms of the "two parents" hypothesis, as birds such as robins, finches, warblers, sparrows, and so forth produce nestlings, which, at hatching, are about as helpless as human infants. Admittedly, they grow more rapidly, and this accelerated growth often requires committed efforts on behalf of at least two adults to provide the ravenous young with a regular stream of nourishment. Mammals, by contrast, are uniquely adapted to be nourished by just one individual—the mother—which, in turn, explains why paternal behavior is rare among mammals. Human beings are something of an exception, at least in part because our infants are so needy as to be somewhat bird like.

There are, however, some examples of mammals that are good fathers, and strong evidence that their paternalism is crucially important. The Djungarian hamster (a species found not, as you might think, along the Danube, but rather the steppes of Mongolia) appears to be monogamous. Male Djungarian hamsters are also notably solicitous and paternal; they assist their mates in the mechanics of giving birth by removing fetal membranes, opening the newborn's nasal passages by licking them, and so forth.[10] Djungarian hamster pups reared in biparental families have a 95% survival rate (by weaning), whereas among those reared by single moms, only 47% survive.[11]

Owl monkeys are among the very few nonhuman primates that appear to be genuinely monogamous. It is noteworthy that these animals (doubly unusual in that they are also nocturnal) have sperm counts that are exceptionally low—because of their monogamy they don't have to compete with the sperm of other males—and their paternal care is very high: given their mating system, a male owl monkey can be confident that he is the father of his mate's offspring.[12] The pattern should be clear: across animal species, monogamy correlates strongly with biparental care, mostly by allaying the father's uncertainty about his genetic patrimony. I haven't seen data on the correlation between men's doubt about paternity and their refusal to provide child support—not to mention divorce rates—but the connection is so predictable that it seems unnecessary to validate, which is probably why no one has bothered to do so!

For Friedrich Engels, monogamy was a straitjacket imposed on women by men, "based upon the supremacy of the man, the express purpose being to produce children of undisputed paternity; such paternity is demanded because these children are later to come into their father's property as his natural heirs."

A system of overt polyandry, by contrast, would necessarily entail paternal uncertainty among the co-husbands, each of whom must settle for a reduced probability of being the genetic father. And of course, in the absence of genetic testing, a recent and—even now—rare thing, men cannot identify their genetic offspring. As a result, polyandrously mated fathers are unlikely to be seriously involved in child care, which in turn rebounds to the disadvantage of the children as well as their mothers. Not surprisingly, therefore, polyandry is hardly ever institutionalized; and, moreover, when it occurs (as in the Tre-ba people of Tibet), it is strongly biased toward fraternal polyandry, by which the co-husbands are at least guaranteed to be uncles if not genetic fathers. When polyandry occurs in other circumstances—which is to say, female infidelity—it is nearly always kept secret by the women who engage in it unless it is accompanied by the useful fiction of "partible paternity."

After my wife and I wrote *The Myth of Monogamy: Fidelity and Infidelity in Animals and People*,[13] we were besieged by members of the "polyamory" community, people who believe in the desirability of what used to be called "free love" and who revel in the notion that monogamy is "unnatural." They were disappointed that we could not validate their chosen lifestyle, not so much because we disapproved as because human biology—along with mammalian biology generally—does so. The problem is twofold: on one hand, even though human beings (like most organisms) are predisposed to have multiple sexual partners, they are far more enthusiastic about their own sexual freedom than that of their partner. Sexual jealousy—of one's mate—is at least as potent as sexual profligacy for one's self. And second, once doubt is raised about paternity, an evolutionary monkey wrench is introduced into the potential for biparental child care.

It is one thing to "stay together for the sake of the children" and quite another to acknowledge that when monogamy occurs in the natural world, it is typically—and often, solely—because of the benefit monogamy confers to those children.

This doesn't mean, however, that both males and females—of many different species—don't attempt to have their biparental cake and eat it too (i.e., to indulge in covert polygyny or polyandry themselves while being intolerant of similar behavior by their partner). Among many species, males are polygynous if they can manage it, accepting monogamy as a consolation prize. Male long-billed marsh wrens, for example, eagerly solicit additional females as soon as they have accomplished a mating. As the breeding season progresses, however, if their extracurricular efforts are in vain, they revert to being diligent fathers and partners in assisting their "first wife" in provisioning the offspring.[14]

Sometimes there is a true battle of the sexes resulting in downright unfairness in favor of whoever wins. Among European birds known as pied flycatchers, both parents typically assist in provisioning the ravenous nestlings. However, males will often attempt to have a "mistress" on the side in addition to their primary mate. In such cases, the philandering male will court this additional female, cooperate in nest building, and mate with her; but the male won't bring any insects to help feed his secondary offspring, devoting all of his paternal assistance to his primary mate. As a result, the secondary female must function as a single parent; and not surprisingly, she rears significantly fewer fledglings than does the primary female, who benefits from the male's foraging efforts. Interestingly, to court a secondary female, the male pied flycatcher establishes a separate territory, some distance from his real "wife" and her nest, suggesting that he is doing so to keep his "mistress" from knowing that he is already bespoken.[15] The research article was titled "The Conflict Between Male Polygamy and Female Monogamy."

Females, too, can be selfishly sexual, faking monogamy but acting polyandrously for their own fitness benefit. Among osprey—also known as fish hawks—a male brings food to his incubating mate. If he doesn't do an adequate job of bringing home the salmon, the female will copulate with another male, who then brings fish to her nest . . . when the cuckolded "husband" osprey is away.

Men and women—like male and female pied flycatchers or osprey—can also differ in their evolutionary interests, despite the fact that human beings are if anything somewhat unusual in often being coshareholders in each other's biological fitness. Accordingly, monogamy is both a consequence of that shared interest and—if sexual monogamy converges with social monogamy—a generator of it.

The connection between monogamy—genuine or faked—and effective parenting could also involve more than simple biparental care. There is, for example, the "bodyguard" hypothesis, which suggests that one-to-one, male–female bonding may have evolved as a means whereby females gained protection from unwanted sexual attention by other males. This idea was first developed with regard to baboons, among which males and females often develop friendships, which subsequently expand to include copulation, after which other males typically refrain from harassing a female who is "going steady" with an identified male.[16] There is every reason to expand the picture to include prehistoric human beings as well.

One problem with this idea, however, is that it isn't clear why females would consider such extracurricular attention unwanted, especially given the likely payoff of confusing paternity and thus gaining protection against infanticide. As to why other males might have respected these budding proto-monogamous pairings, the body-guard hypothesis has an answer: it may have been a good deal if male–male cooperation is sufficiently important: that is, if males, whose bonds with a particular female are respected, would have been more likely, in return, to cooperate in coalitions with other males, as well as showing reciprocal respect for the monogamous pairings of those other males who play by the rules. This isn't very far from other proposals basing monogamy on a kind of egalitarianism (more on this later in the chapter).

Another related possibility is that males were selected to associate with females to protect the females and their shared offspring, not so much from unwanted sexual attention by other males but from infanticide itself—that is, to protect their genetic investment.[17] It would be nice if the lure of positive male care was the major reason, but not impossible that paternal defense against some very negative male infanticide was at least as important.

"Among our hominid ancestors, a simple security arrangement between the sexes would have been easy to expand upon," writes primatologist Frans de Waal:

> For example, the father could have helped his companion in locating fruit trees, capturing prey, or transporting juveniles. He himself may have benefited from her talent for precision tool use, and the gathering of nuts and berries. The female, in turn, may have begun prolonging her sexual receptivity so as to keep her protector from abandoning her. The more both parties became committed to this arrangement, the higher the stakes. From an evolutionary perspective, investment in someone else's progeny is a waste of energy; therefore, males may have tightened control over their mate's reproduction in direct proportion to their assistance to her.[18]

It is also possible that human monogamy was especially important in the context of infant vulnerability, with a predictable dropping off in passion and mutual commitment as offspring become more independent. A study of 58 different human societies found that divorce rates tend to peak at about four years after the birth of a child,[19] suggesting that the primary function of monogamous unions is to get childrearing off to a good start, after which the parents are inclined to start over, not uncommonly with someone new. Of course, it's one thing for a child to survive to weaning and quite another for him or her to be counted an evolutionary success—that is, to reach reproductive age. But such findings as the "four-year itch" make a strong case for "natural" serial monogamy (a close cousin to polygamy) in human beings.

There are at least two other factors related to parenting but not strictly connected to confidence of paternity that might also contribute to human monogamy. One is the so-called demographic transition, a well-known and widespread phenomenon, whereby people consistently reduce their family size in proportion as they experience improved medical care, economic development, and especially improvement in the social status and education levels of women. Ecologists have long identified a continuum of reproductive strategies, ranging from "r-selection" at one end to "K-selection" at the other. Think of the difference between mice (r-selected) and elephants (K-selected). The "r" in r-selection refers to the natural rate of population increase, and r-selected creatures tend to be small-bodied, rapidly growing, early maturing animals that are essentially adapted to live fast, love hard, and die young. By contrast, K-selected organisms are likely to be large-bodied, slow growing, late maturing, and inclined to breed at a more deliberate pace. The "K" in K-selection derives from the German word for carrying capacity because K-selected creatures typically live at or near their environment's ability to sustain a given population.

If you are a species that regularly experiences conditions in which there is an opportunity to breed rapidly and fill up an otherwise unoccupied niche, r-selection is your ticket. Think of how mouse populations can almost literally explode. Under an r-selection regime, there is little payoff to being especially well nourished, educated, or protected by your parents; the key is to make lots of kids and do so quickly. On the other hand, if you live in an environment that is already at or near the ecological carrying capacity for your species, the chances are that simply focusing on quantity of offspring will be less successful in the long run than if you were to emphasize quality. And so, K-selected animals grow slowly, produce fewer offspring at a time and with a longer delay between repeated egg laying or giving birth, and they are likely to lavish substantial parental care and protection on each one.

Where do human beings fit on the r- and K-continuum? Clearly toward the K end. But it is clear that the human reproductive strategy isn't carved in stone. People become even "less r" and "more K" as modernization takes hold. Demography, the study of human population, is a notoriously squishy science, subject to any number of exceptions and unpredictable variations. Amid the chaos, however, this pattern persists: cross-culturally, when socioeconomic conditions improve and societies modernize (notably including higher and more widely shared education levels, medical care, and increased status for women so that their perceived value is no longer determined primarily by their fecundity), the intervals between births increase, along with age at first reproduction; overall, birth rates decline. Sometimes they even plummet, all without any direct antinatalist government intervention. It appears that the demographic transition is another "cross-cultural universal," a predictable consequence of human nature under certain circumstances.

Here are some examples. The overall fertility rate in India fell significantly (despite the loosening of Indira Gandhi's draconian population policy), from 3.6 in 1991 to 2.3 in 2013. Nevertheless, a substantial urban–rural divide remains in that country, with village women having a fertility rate of 2.5 children compared to urban residents' 1.8. According to estimates from the CIA World Factbook, covering 224 countries in 2014, the half dozen with the highest total fertility rates (the average number of births per woman in the population) were, in order, Niger, Mali, Burundi, Somalia, Uganda, and Burkina Faso, with numbers running from 6.89 to 5.93. At the opposite end, Singapore had the lowest total fertility rate (0.80), preceded by Macau, Taiwan, Hong Kong, South Korea, and the British Virgin Islands (1.25). The United States was 123rd (2.01), roughly between Ireland (2.00) and New Zealand (2.05). Get the picture?

Further evidence (much more is available) comes from the United Nations Human Development Report for 2013, which makes use of the Human Development Index for each country, a composite statistic originally created by the economists Amartya Sen and Mahbub ul Haq, which combines data for income, education level, and life expectancy. Dividing countries into four groups with regard to their human development—very high, high, medium, and low—the report shows a striking inverse relationship with fertility rates: the greater the human development, the lower the fertility rate.

Most of the developed, industrialized world has virtually attained zero population growth, whereas Africa and parts of the Islamic world retain the highest birth rates. In nonindustrial, rural societies experiencing the first stage of this demographic process, birth rates and death rates both tend to be high, and the population therefore remains relatively stable. Then, with public-health measures such as immunizations, widespread food distribution, and so on, death rates decline while birth rates remain high, and the population therefore increases. But in the final stage of the demographic transition, improving social and economic conditions, along with lowered infant mortality, produce a desire for smaller families, as parents realize they do not need large numbers of children to serve as field hands, compensate for those they might lose, or provide social security in their old age. Moreover, parents recognize that to endow their children with benefits such as a higher education, they must have fewer of them. As a result, populations eventually level off.

Although the demographic transition is predictable, it isn't simple. In some ways, it is counterintuitive, especially from a Darwinian perspective. Thus we might expect that healthier, wealthier people would transfer their bounty into *more* children because offspring constitute the evolutionary bottom line for all living creatures. But that is where things get especially interesting.

I suggest that the demographic transition reflects a transition from a predominantly r-strategy to a K-strategy. Does this mean that what demographers tell us happens in human societies—the demographic transition—is literally the same process biologists describe in other organisms? Clearly there is a convergence in outcome. But is it based on shared evolutionary pressures and manifested by comparable genetically encoded tendencies or just a notable coincidence? Frankly, no one knows.

Natural selection has placed different species in different locations on the r- and K-continuum; and as far as we can tell, most of them are severely limited in their ability to make adjustments, even when conditions change and especially if they change a lot. This is because evolution, in itself, doesn't have any foresight. But—and here is the key—natural selection can (at least on occasion) create organisms that do have it, or at least the ability to assess circumstances and respond accordingly: for example, us. A mouse cannot suddenly revise its entire biology in response to new circumstances. It can no more become a K-selected pachyderm overnight than an elephant can instantly become an r-selected rodent. People, too, cannot fundamentally redesign their anatomy or physiology, but they can adjust their reproductive behavior, recognizing that when it's time to focus on quality rather than quantity, it's appropriate to have fewer children and invest more in each one.

We do all sorts of things that demonstrate foresight, such as contributing to a child's college tuition account or storing and then planting seeds that won't germinate until the next season. On the other hand, when it comes to some of the most important situations in which foresight appears especially needed—to intervene in our current path toward global climate disaster, counter reductions in biodiversity, take bold steps toward eliminating the risk of nuclear war—our species has been woefully shortsighted. But maybe, just maybe, when it comes to the intimate decision of family planning, people are somewhat far-seeing, at least when it comes to their enlightened self-interest. And if so, it is at least possible that monogamy is part of the equation because as we have seen, it is precisely under monogamy that biparental care is most likely, and such care is exactly what is needed if parents are to have fewer children and take better care of each.

Here is the other hypothesis that might connect monogamy with parenting. It turns out that monogamy is more frequent in environments that are severe and in which husband–wife cooperation is important—and sometimes necessary—for the family unit to obtain enough food and protection from environmental threats: for example, the high arctic. Under more benign circumstances, polygyny is more likely to flourish.[20] On a related note, a cross-cultural survey concluded that as the male contribution to food production increases, the frequency of polygyny decreases and monogamy increases.[21]

A cynic might note that monogamy often breeds monotony, to which an evolutionist could counter that "breeding" itself is a potent goal, and one at which monogamy excels.

Next is the connection between monogamy and equality, as well as its linkage to social and domestic peace. Friedrich Engels had little patience with monogamy, seeing it as the institutionalization of oppression:

The first class antagonism which appears in history coincides with the development of the antagonism between man and woman in monogamous marriage, and the first class oppression with that of the female sex by the male. Monogamy was a great historical advance, but at the same time it inaugurated, along with slavery and private wealth, that epoch, lasting until today, in which every advance is likewise a relative regression, in which the well-being and development of the one group are attained by the misery and repression of the other. It is the cellular form of civilized society, in which we can study the nature of the antagonisms and contradictions which develop fully in the latter.[22]

Engels, like his more famous collaborator, Karl Marx, isn't terribly popular these days. Nevertheless, he was right to identify the "antagonisms and contradictions" that often come along with monogamous marriage, plus the reality that overwhelmingly, it is men oppressing women. Here is Engels, again:

> It was the first form of the family to be based, not on natural, but on economic conditions—on the victory of private property over primitive, natural communal property. The Greeks themselves put the matter quite frankly: the sole exclusive aims of monogamous marriage were to make the man supreme in the family, and to propagate, as the future heirs to his wealth, children indisputably his own. . . . Thus when monogamous marriage first makes its appearance in history, it is not as the reconciliation of man and woman, still less as the highest form of such a reconciliation. Quite the contrary. Monogamous marriage comes on the scene as the subjugation of the one sex by the other; it announces a struggle between the sexes unknown throughout the whole previous prehistoric period.

Engels was wrong, however, to lay all the blame at monogamy's door: as already explained, there is reason to conclude that if anything, polygyny is more oppressive of women (and even more so of most men), and that monogamy—troublesome as it often is—represents an improvement: or at least, a movement toward greater equality.

It may well be that—to paraphrase Winston Churchill's observation about democracy—monogamy is the worst of all possible arrangements . . . until you consider the various alternatives!

At the same time, monogamy is hardly a cure for sexism and the oppression of women. Modern Thailand, for example, is socially monogamous but also rigidly patriarchal, with wives having little power. There is a long-standing tradition of married Thai men visiting prostitutes, as well as, for some, maintaining "minor wives" who receive some financial support. Thai wives overwhelmingly prefer their husbands—if they are going to be sexually active outside marriage—to do so with prostitutes than to maintain a "minor wife," simply because professional sex workers siphon off less resources and typically are less threatening to the primary marriage. With the advent of AIDS, interestingly, some of this preference has been shifting, because regular sex with a "minor wife" is less risky than with a seriously polyandrous professional.[23]

With genuine honest-to-Darwin monogamy, the evolutionary interests of man and woman are identical (at least when it comes to their coproduced offspring). One payoff to monogamy, therefore, would be that it reduces conflict within each household. After all, if philandering is ruled out, couples would have a whole lot less to worry, fight, and divorce about. It is less clear whether strict and genuine monogamy would also reduce competition among women. Arguing for diminished interwoman competition under initial monogamy is that women should be less prone to resent the addition of a new and younger wife to join that of a wealthy and powerful man; they would also be spared the aggressive competition among co-wives that at least sometimes results in higher offspring mortality (as described for the Dogon in chapter 4).

On the other hand, if men differ substantially in their desirability as mates, then because under monogamy even the most desirable men end up with just one wife, it might if anything result in greater woman–woman competition to be that one. We've already seen that under polygyny, competition among co-wives can be substantial, just as dowry payments may reflect woman–woman (and family–family) competition in cases of monogamy.

Monogamy is maintained among a number of animal species by the intensity of female–female competition—specifically, by the aversion that an existing, monogamously paired "wife" shows toward a possible second wife. It is also true, interestingly, that among many animal species, females enforce monogamy,[24] sometimes using techniques that go beyond simple aggression. Among starlings, for example, mated females solicit additional copulations with their male partner if he begins showing interest in other females.[25]

Monogamy in the modern Western world is socially imposed, mandated by law as well as by religious demands. But why? What is society's payoff here? As noted previously, one key is widespread recognition of the benefits of biparental childrearing. After all, deadbeat dads aren't just socially frowned on; in most cases, they really are bad for the children in question as well as for the children's' mothers. Equally important, however—at least at society's macro level—is the benefit provided by monogamy's role in damping down male–male competition.

It is a cross-cultural universal that men—especially young men—are the biggest troublemakers, jostling with each other and with the existing power structures, if only because (whether they recognize it or not) they are selected to seek their own measure of reproductive success. As zoologist Richard Alexander has written

> It is not trivial that socially imposed monogamy . . . correlates with (1) justice touted as equality of opportunity; (2) the concept of a single, impartial god for all people; and (3) large, cohesive, modern nations that wage wars and conduct defense with their pools of young men. To a large extent socially imposed monogamy has spread around the world by conquest. The social imposition of monogamy thus simultaneously (1) inhibits the generation of certain kinds of within-group power dynasties that might compete with government and lead to divisive within-group competition and (2) promotes those activities and attitudes that generate and maintain success in the wielding of reciprocity as the binding cement of social structure (honesty, sincerity, trust).[26]

Another cross-cultural universal is the very high representation of young men among warriors, whether recruited, coerced, or volunteered. This is frequently due not only to preferential enlistment by the young warriors themselves (seeking, among other things, to enhance their eventual social and breeding prospects) but also to a preference on the part of the society's old men. And this, in turn, likely derives in part from a realization that young men (and women) are more readily manipulated by the demands of their elders; hence, they make better soldiers than do older recruits or conscripts. In addition, of course, young men are more prone to think of themselves as potentially immortal. Being young, they are also likely to be in better physical condition and thus better fighters. But we shouldn't overlook the fact that in proportion, as young men are siphoned off via the battlefield, there are fewer of them making trouble back home.

Elders in positions of power are faced with a quandary: on one hand, they are threatened by each upcoming generation of young men whose troublesome penchant for risk-taking just might, on occasion, pay off for the risk-takers. On the other hand, elders often want to encourage precisely such risk-taking, or at least to channel it toward defense of the larger social unit. Both these problems could be solved, or at least ameliorated, by institutionalizing monogamy, so that young warriors (if they survive) have domesticity awaiting them when they return. Besides, why should a warrior risk death on the battlefield to defend someone else's harem?

Continuing in this vein, how do warriors from monogamous societies compare with polygynists? There are currently no data that speak to this question. It seems likely, however, that a soldier with one or more children (more frequent in monogamous than in polygynous societies) would be more cautious, although not necessarily less brave, than someone who has no young dependents. On the other hand, men who have already obtained mates, or have the expectation of doing so after a war, might be not only more resistant to forced conscription, but more disposed to desert or malinger because of worry about the fidelity of their wives or girlfriends back home. Among the propaganda maneuvers broadcast by the infamous Tokyo Rose during World War II was the leering insinuation that while he was fighting in the Pacific, G. I. Joe's girlfriend was being unfaithful. But perhaps such anxiety—more characteristic of a monogamous society in which each man might have a wife or girl friend—is less socially debilitating than having few sociosexual prospects at all, which is more likely under polygyny.

It could also be that soldiers from polygynous societies—having fewer options for domestic bliss at home—would be more prone to acts of self-sacrifice and more inclined to act so as to make themselves stand out (via promotions, medals, etc.) in a way that military elders would approve. After all, so long as enough of their junior competitors get killed or maimed, male–male competition is reduced at home, even if a few heroes return to public acclaim as well as to literally welcoming arms and warm beds. These questions could presumably be settled by empirical research.

In any event, and for whatever reason, it remains true that even as huge nations (China now, India in the recent past) attempt to control their populations, no country can get away with prohibiting its citizens the right to reproduce.[27] The cultural imposition of monogamy isn't alone when it comes to a kind of societal leveling, at least in theory. Equalizing opportunity seems to be one of the consistent trajectories of most modern and modernizing countries: improving the status and opportunities of women, of racial and ethnic minorities, guaranteeing one person-one vote, providing increased access to people with disabilities, and so forth.

It probably isn't coincidental that Christianity—which Nietzsche sneeringly derided as a "slave morality" because of the aid and comfort it gave to the downtrodden—has been especially active in promoting monogamy: "A man will leave his father and mother and be joined to his wife, and the two shall become one flesh. . ... So then, they are no longer two but one flesh. Therefore what God has joined together, let not man separate" (New American Standard Bible, Matthew 19: 4–6). Not to mention the more libidinous Proverbs 5:18–19: "Rejoice with the wife of thy youth . . . loving hind and pleasant roe; let her breasts satisfy thee at all times; and be thou ravished always with her love." Unspoken here is the pious hope and expectation that the more satisfied the husband, the less trouble will he make.

Anthropologist Bernard Chapais has made an intriguing suggestion along these lines. Instead of seeing monogamy as a mechanism to diminish the disruptive potential of unmated, unsatisfied males, Chapais hypothesizes that the advent of handheld weapons early in human evolution would have served—a bit like the Colt .45 in the American West—as a kind of equalizer, making male–male competition more balanced as well as potentially more lethal.[28] The result could have been that instead of retaining their harems, increasing numbers of otherwise dominant men would have found it in their interest to "allow" other men to have their own wives. If so, then greater equality (resulting from widespread weaponry) would have been a driver of monogamy rather than a targeted consequence.

All of this fits quite nicely with our earlier observations that human polygyny is unavoidably a situation of conflict, with polygynist men gaining and polygynist women typically losing; and with unsuccessful, unmated men being the biggest losers. Hence, the transition from polygyny to socially imposed monogamy is one from conflict-ridden special privilege to more widely shared benefit.

Elitism, like submission, has a nasty connotation, even as it exerts enormous appeal, notably to those who see themselves as either members of the elite or who fall prey to the overly optimistic (and notably American) presumption that any day now they will enter the blessed club of wealth plus sexual privilege. Until then, the leveling effects of social pressure certainly haven't been universal, or even widely shared. Witness the grotesque increase in income inequality within the United States, but also the growing outrage it has provoked. It seems likely, however, that any spread of democracy worldwide will be associated with a decline in sanctioned polygyny, a system that should naturally become less appealing—eventually becoming downright untenable—in proportion as something approaching socioeconomic equality is reached: if ever.

I've already noted that Indonesia is the largest country (by population) in which polygyny is legal. Here, polygynous politicians are notably unpopular. They are especially opposed by women voters, but are also increasingly resented by their male constituents as well—although this is rarely expressed, maybe because men harbor stubborn but forlorn hopes of eventually joining the promised land of upper echelon polygynists. In addition, even among countries in which polygyny is legally outlawed and monogamy mandated, it is pretty much guaranteed that many (perhaps most) powerful and wealthy men will be de facto polygynists, either via numerous sequential marriages or by having multiple mistresses. But this leaves society, once more, with a surfeit of troubled and troublesome unmated young men. Or at least, jealous of others' success.

When it comes to the potential payoff to society that comes from pacifying those restless and horny male elephant seals who are more sexually frustrated in proportion as the social system is highly polygynous, I cannot do better than to share once again an extended quote from the garrulous, cynical H. L. Mencken, whose prose harks back to a more relaxed era of verbal pyrotechnics. Writing of monogamy as a pacifying institution, Mencken opined that

> Civilized men are in favour of it because they find that it works. And why does it work? Because it is the most effective of all available antidotes to the alarms and terrors of passion. Monogamy, in brief, kills passion—and passion is the most dangerous of all the surviving enemies to what we call civilization, which is based upon order, decorum, restraint, formality, industry, regimentation. The civilized man—the ideal civilized man—is simply one who never sacrifices the common security to his private passions. He reaches perfection when he even ceases to love passionately—when he reduces the most profound of all his instinctive experience from the level of an ecstasy to the level of a mere device for replenishing armies and workshops of the world, keeping clothes in repair, reducing the infant death-rate, providing enough tenants for every landlord, and making it possible for the Polizei to know where every citizen is at any hour of the day or night. Monogamy accomplishes this, not by producing satiety, but by destroying appetite. It makes passion formal and uninspiring, and so gradually kills it.

Mencken went on, and I cannot resist sharing his scrumptious, albeit excessive elaboration:

> The advocates of monogamy, deceived by its moral overtones, fail to get all the advantage out of it that is in it. Consider, for example, the important moral business of safeguarding the virtue of the unmarried—that is, of the still passionate. The present plan in dealing, say, with a young man of twenty, is to surround him with scare-crows and prohibitions—to try to convince him logically that passion is dangerous. This is both supererogation and imbecility—supererogation because he already knows that it is dangerous, and imbecility because it is quite impossible to kill a passion by arguing against it. The way to kill it is to give it rein under unfavourable and dispiriting conditions—to bring it down, by slow stages, to the estate of an absurdity and a horror. How much more, then, could be accomplished if the wild young man were forbidden polygamy, before marriage, but permitted monogamy! The prohibition in this case would be relatively easy to enforce, instead of impossible, as in the other. Curiosity would be satisfied; nature would get out of her cage; even romance would get an inning. Ninety-nine young men out of a hundred would submit, if only because it would be much easier to submit than to resist.

It is easy to discount Mencken's scoffing tone, and yet the world witnessed the potency of frustrated bachelor elephant seals during the Arab Spring uprisings of 2010. Although little noted by the Western press, a substantial underlying driver of social unrest was the widespread tradition in many Arab societies whereby even monogamous marriage isn't tenable until a would-be groom has accumulated sufficient wealth to "afford" a bride.

A similarly troublesome phenomenon is particularly afoot in modern Asia as well, where vigorous government efforts to restrict family size—notably in China and India—have resulted in substantial sex ratio imbalances, leading to an excess of men. (This is because in many traditional societies, males are valued over females, so that when a family is limited to one or at most two children, female infants are more liable to be aborted or killed shortly after birth.) The resulting excess of men isn't itself a direct consequence of polygyny, but it has a disruptive social effect that parallels and reinforces that of polygyny itself: predisposing societies toward social disruption, increased violence, and possibly even war, ultimately because the reproductive prospects of these "excess" men are limited or even foreclosed altogether.[29] We need more attention to the societal consequences of altered sex ratios, whether generated by differential mortality or by social traditions—both of which in turn are often driven by our polygamous ancestry.

Even the seemingly strait-laced Catholic church has long acknowledged the need to provide acceptable sexual outlet, particularly for young men; and going beyond Paul's famous exhortation that although marriage is somehow regrettable, it is "better to marry than to burn." Thus, Augustine even supported prostitution, noting that "remove [it] from human affairs and you will unsettle everything because of lust." A millennium later, Aquinas was somewhat earthier, comparing prostitution to a "sewer system in a palace. Do away with it and the palace will become a place of filth and stink."

These are strong words, and doubtless exaggerated. But if nothing else, they give credence to those who claim that monogamy itself is merely a form of enforced prostitution of women in the service of pacification of men. Fortunately, there is another perspective.

Alert readers may have noticed that "love" hasn't yet appeared in this book— but not because it isn't real or relevant. It is genuinely mysterious whether love can coexist with polygamy, for either men or women. Several things, however, aren't mysterious at all. For one, even though our species is clearly primed for polygamy, we are *also* predisposed to pair-bonding because from evolution's perspective, multiple parental figures are better than one (and much better than none); and also, most human beings, through most of our ancestral history, have probably been monogamously mated, if not by choice, then simply because of circumstance. And so, an objective, albeit cynical take on love is that it is a mechanism facilitating a deep personal connection when a multitude of such connections aren't feasible.

Some people would doubtless not acknowledge their own polygamous temptations, and would in any event object to objectifying so seemingly magical a sensation as romantic love, worried that by doing so they drain it of romance and of a certain ineffable appeal. In his poem, *Lamia*, John Keats remonstrated with Newton for explaining the optical physics of rainbows, thereby draining them of magic and poetry:

> Do not all charms fly
> At the mere touch of cold philosophy?
> There was an awful rainbow once in heaven:
> We know her woof, her texture; she is given
> In the dull catalogue of common things.
> Philosophy will clip an Angel's wings,
> Conquer all mysteries by rule and line,
> Empty the haunted air, and gnomèd mine—
> Unweave a rainbow

I would hope that knowing something about the moon's structure—and deleting both the man in the moon and any supposed green cheese—hasn't diminished the joy of a moonlit night, just as understanding that a rainbow is formed by the bending of light by droplets of water only "unweaves" our ignorance, leaving the rainbow no less beautiful:[30] and if anything, more so, for being better known.

Nonetheless, maybe ignorance really is bliss, at least for some people. "Self-contemplation is a curse," wrote the poet Theodore Roethke, "that makes an old confusion worse." Self-contemplation has nonetheless long been a human preoccupation, and self-knowledge—especially when combined with other-understanding—can be no less rewarding than comprehending rainbows, moonlight, why the kangaroo has a pouch, or why certain organisms are prone to bonding with others of their species.

As for love, there are, of course, many forms, including love of parent for child and vice versa; love of country or of a certain food, color, weather pattern or even a sports team; love of life itself, of God for those so inclined, of humanity in general, or of someone in particular: different details, but very similar underpinnings, involving a strong commitment and connection, in many cases carrying over into willingness to sacrifice one's self for the beloved. When it comes to a hardheaded (and to some people, perverse) biological perspective, the romantic, bonding love of adults is a tactic whereby evolution gets people to commit strongly and deeply to each other because the ultimate adaptive value of such involvement exceeds the payoff of remaining unattached, unloving, and unloved.

In Plato's *Symposium*, Aristophanes presented a "theory" for the origin of human bonding. It seems that long ago there were no people as we know them today but rather Androgynes, large-bodied creatures with two heads, four arms and legs, and multiplied genitals of a diversity and arrangement that can scarcely be imagined. Great was their power and ambition, such that Zeus felt a need to take them down a notch or two (in fact, precisely two), so he belabored them with thunderbolts as was his wont in those days, splitting each Androgyne in two. Since then, their descendants have been reduced to wandering the Earth in a state of less-than-androgynous incompleteness, each seeking for his or her missing half.

For better or worse, we're not like that. Long-term pair bonds are universal to human societies, coexisting uneasily with polygamous penchants and a susceptibility to sexual opportunism. Moreover, as described in the next chapter, our situation is more complicated yet because although what's sauce for the human goose is equally saucy for the gander, each sex prefers—once again, for entirely understandable, adaptive reasons—that it be given more lascivious leeway than its partner.

Nonetheless, maybe the origin of monogamy is as simple as *amor vincit omnia*: love conquers all. Really. Can romantic love also coexist with multiple sexual relationships? It certainly seems more compatible with monogamy than with polygamy (or polyamory); and it could be that despite the all-too-human penchant for polygamy, one thing that not only permits monogamy and makes it viable but actually provides a positive push is old fashioned romantic love. This is not the place to elaborate on love's slings and arrows, delights and glories, except to note that a cross-cultural review of 166 societies noted that romantic love was identified in 151.[31] There is no doubt that romantic love is a cross-cultural human universal; a doctoral thesis (Human nature and the nature of romantic love), reviewed data, and accounts from 100 different human cultures finding clear evidence of romantic love in every case.[32] Evidence from the Human Relations Area File confirms pair-bonded romantic love in fully 89% of all described cultures.[33]

Historian and cultural philosopher Denis de Rougemont disagrees with the view that equates monogamy to monotony, that proclaims monogamous marriage to be "the grave of love." He says that it is, rather, the "grave of *savage* love," and that

> Savage and natural love is manifested in rape . . . But rape, like polygamy, is also an indication that men are not yet in a stage to apprehend the presence of an actual person in a woman. This is as much as to say that they do not know how to love. Rape and polygamy deprive a woman of her equality by reducing her to sex. Savage love empties human relations of personality.[34]

De Rougemont goes on to argue that

> A man does not control himself owing to a lack of passion . . . but precisely because he loves and, in virtue of his love, will not inflict himself [on others]. He refuses to commit an act of violence which would be the denial and destruction of the person. He thus indicates that his dearest wish is for the other's good.

According to anthropologist Helen Fisher, we have three different—yet related—love systems: lust (directed to mating and the mechanics of reproduction), romantic attraction, and regular old, garden variety attachment.[35] A number of other distinctions have been proposed, usually involving separate categories for passionate love (sexual desire) and some form of "companionate love," which calls for commitment and deep attachment, but with fewer hormones, bells, and whistles.[36] The likelihood is that the former is generated by natural selection as a means of achieving fertilization—either getting fertilized or fertilizing others, depending on one's sex—while the latter is ultimately buttressed by the benefits of shared parenting.

As Goethe reminded us, "Love is an ideal thing, marriage a real thing; a confusion of the real with the ideal never goes unpunished." At the same time, real love and marriage are not only compatible, but just as love can lead to marriage, marriage can lead to love. The remarkable success rates of arranged marriages, around the world, attests to this.

I don't know how many roads actually lead to Rome, but biologists have identified many independent routes whereby marriage can lead to love. In an earlier book, my wife and I described some of the biological factors that mitigate polygamy and make human monogamy at least possible and in many cases, desirable.[37] These include, in addition to various evolutionary considerations, a quartet of mechanisms that support monogamy, together constituting what might be called the Four Horsemen of the Monogamist. First is "attachment theory," initially propounded by the British child psychiatrist John Bowlby.

It emphasizes that human infants and their parents connect automatically and deeply, from a very early age, serving needs that go beyond mere sustenance. Attachment theory was initially developed to explain parent–offspring love, but it seems likely to have implications for monogamy as well, constituting what biologists call a proximate mechanism, that is to say, something that helps explain the *how* of behavior, rather than the deeper *why*.[38] More recently, scientists have also identified three additional proximate mechanisms, all operating via neurobiology, by which social attachment seems to take place: mirror neurons, neuroplasticity, and certain hormones, especially the endogenous "love-drugs" oxytocin and vasopressin.

Let's look at each of these, briefly. Bowlby italicized the crucial role of infant–mother attachment, not only for normal emotional development but for basic mental and even physical health. (Incidentally, this work made Bowlby the most-cited psychologist of the early 21st century, exceeding Freud, Jung, Piaget, or Skinner.) He began his research during World War II, studying the reactions of babies who were separated from their mothers when hundreds of thousands of young children were evacuated from London due to the *Luftwaffe's* bombing campaign. Although well cared for materially, these infants often went through predictable phases of mourning. Although some rebounded quite well, others fell into a deeply depressed state known as "anaclitic depression" or marasmus. They literally turned to the walls, and quit eating and drinking. Some of them died.

According to Bowlby, basic attachment starts with every infant's need for secure connection to an adult caregiver, especially during a crucial period from birth to two years of age, and particularly during times of stress—but not only then. Bowlby emphasized that the absence of a responsive, sensitive adult is itself a major cause of stress and that "separation anxiety" follows loss of such an attachment figure.

For our purposes, attachment theory suggests that the human behavioral system is predisposed to form exactly the kind of deep interpersonal connection that monogamy also requires; and, moreover, that an adult–adult bond may well be part of a healthy continuum involving attachment generally, from infancy to old age. Bowlby coined the word "monotropy" to describe what he saw as a child's bias toward attaching primarily to one caregiving adult figure. That term is now defunct, as developmental psychologists have become convinced that infants are not, shall we say, monogamous, in their early attachments.

I suggest that Bowlby was nonetheless more correct than even he realized and that human beings are wired for attachment. Not only is it likely that one's early attachment experience—for good or ill—has repercussions when it comes to subsequent adult bonding, but attachment itself is a basic part of being human, at any age. As Bowlby put it, "Attachment theory regards the propensity to make intimate emotional bonds to particular individuals as a basic component of human nature, already present in germinal form in the neonate and continuing through adult life into old age."[39]

Turning now from attachment to detachment, or at least, the formation of new connections, it had long been thought that nerve cells were incapable of new growth: you can't teach an old dog new tricks, and all that. Brains—especially adult brains—were assumed to be hardwired, incapable of significant change. This is no longer valid. The dogma that brains, unlike, say, skin, couldn't repair themselves or establish new connections has been convincingly overthrown in what is perhaps the most stunning discovery by neurobiologists in the late 20th century: neural plasticity.

This simply means that neurons are constantly molded, modified, and strengthened by experience. Specifically, dendritic spines (the microscopic outgrowths on the receiving ends of nerve cells that resemble complex tree branches) sprout in response to repeated stimulation; axons (the long cable by which neurons send outgoing signals) grow; synapses (the tiny spaces between sending and receiving cells) can be modified to facilitate or to inhibit the transfer of information; just as genes within nerve cells are switched on and off in response to stimulation; and—even more striking for those of us trained in the earlier dogma of neural immutability—"neurogenesis" is real: new brain cells can and do grow.

Not surprisingly, most of the research associated with neuroplasticity has focused on its clinical applications, notably ways to facilitate recovery from strokes and other brain damage. For people who have suffered brain and spinal cord injury, it is especially exciting that connections between existing neurons can be altered: not just changes in receptor sensitivity, but also new growth on the part of existing neurons.

Most dramatic of all is neurogenesis, the actual birth of new nerve cells, demonstrated in at least certain brain regions: the hippocampus, olfactory bulb, and cerebellum. It remains to be seen whether neurogenesis occurs elsewhere in the nervous system and whether stem cells can generate significant recovery of function. There is, of course, considerable interest in the specific cell mechanisms—neurochemical, neurogenetic, neuroanatomic, neuroelectrical, and so forth—that generate plasticity. For our purposes, however, the precise details are less important than the bottom line: the fact of neuroplasticity itself, whose existence is no longer in doubt.

Accordingly, the prospect beckons that just as practice playing the guitar develops particular brain regions,[40] practice interacting with one's beloved can develop brain regions that promote attachment à la Bowlby: in short, love. Plasticity also yields a biological counter to the tired clichés about "genetic determinism." Genes and their behavior-generating devices, neurons, are not straitjackets but rather opportunity providers that give people the tools to grow, develop, change, and adapt to new circumstances, including—I strongly suspect—interpersonal attachment. The ability to give up bad habits, and to respond positively to good ones—not to mention good relationships—isn't just a pious, pie-in-the-sky hope but a likely function of hard-nosed, utterly materialist neurobiology. Insofar as a biological take on monogamy builds on neuroplasticity, rather than being constrained by genetics, the human capacity for monogamy is likely empowered by it.

As with neuroplasticity, mirror neurons—the third neuronal facilitator of monogamy—were discovered only recently; and their significance, although murky, is nonetheless suggestive. It has yet to be investigated whether mirror neurons relate to bonding in general and to monogamy in particular—although they probably do. Here is the mirror neuron story, in brief. Neurobiologist Gacommo Rizzolatti at the University of Parma, in Italy, was leading a research team investigating the prefrontal cortex of rhesus macaque monkeys. It had been known that certain motor neurons, responsible for generating movement, fired when the monkeys performed a particular action, such as grasping a banana. Quite by accident, however, Rizzolatti's group found that these same neurons would also fire when the monkey saw *someone else* grasp a banana. So certain neurons didn't only control movement; they responded when the animals perceived *another* individual engaged in a given movement.

These cells, dubbed "mirror neurons," have been reported in human beings, too,[41] although their exact function has been disputed.[42] Some exciting possibilities can nonetheless be glimpsed. Consider empathy, for example, the experience whereby we can "feel someone else's pain," or when we imagine (better yet, know) what it would be like to "walk a mile" in someone's shoes. The prospect beckons that mirror neurons are the biological basis for "intersubjectivity," whereby people (and presumably, certain animals, too) connect their own selfhood with that of others.

Neuroscientist V. S. Ramachandran refers to "empathy" or "Dalai Lama neurons." Could they also be "attachment neurons," "bonding neurons," even "monogamy neurons"? It certainly appears that human beings, along with many other species, are wired to respond to others in a manner that personalizes their experience as one's own. It isn't clear whether people are endowed with especially vibrant mirror neuron systems, in part because ethical restraints on human experimentation make it very difficult to conduct the necessary studies. But the intriguing possibility exists that empathy—along with basic social propriety, which builds on a personal representation of how others are responding to one's own actions—originates in these neurons. It wouldn't be surprising if sociopaths, who are notably deficient in empathy, are also deficient in either the abundance or responsiveness of their mirror neurons.

Let's go a bit further and posit that these neurons not only mirror another person's experience but are also specific with regard to the particular individual whose experience is being mirrored. Mirror neurons fire, for example, when we see a stranger hit his thumb with a hammer; might they not fire even more readily, more persistently, or with more nerve cells being recruited if that "someone else" isn't a stranger but somebody already known? And who is better known than one's mate?

In Tennessee Williams' *A Streetcar Named Desire*, the pathetic Blanche DuBois famously notes as she is being led away to a mental hospital that she has "always relied on the kindness of strangers." Many of us—perhaps most—are moved by the distress of strangers, including that of the fictional Ms. DuBois. But aren't we even more moved by the distress of those we know and love? And if so, what about also being moved by shared joys, accomplishments, and experiences?

There is, accordingly, this possibility, maybe even a probability: that long-term relationships facilitate a greater degree of empathy at least in part because they set the stage for a greater mirror neuron response. If so, then those rare species that appear to be truly monogamous (California mice, Malagasy giant jumping rats, pygmy marmosets, etc.) can be predicted to have more mirror neurons, or more active ones, than do promiscuous or polygynous species. Interestingly, there is evidence (admittedly incomplete and sometimes contradictory) that species predisposed toward pair-bonded monogamy have larger brains than their polygamously or promiscuously inclined colleagues.[43] This may be due to the requirements of dealing with an "intimate other," the necessity of making frequent, prompt adjustments with regard to another adult. But it could also be argued that dealing with lots of others is no less stressful, and equally demanding of brainpower, which leaves the possibility that if monogamists are brainier, maybe this is because they are better endowed with the necessary mirror neurons.

Conversely, those same mirror neurons might also predispose toward monogamy by making one less likely to inflict pain on another, with disinclination varying directly with the extent to which one can literally feel that pain. It is quite clear that "cheating," for example, inflicts emotional pain on one's partner. The more you can feel another's pain, perhaps the less likely you are to cheat.

As for different individuals within the same species, monogamous individuals might be endowed with more mirror neurons—or more active ones—than those who have multiple relationships. If so, would enhanced mirror neuron density or responsiveness be an effect of monogamy, or a cause?

It seems at least possible that having a more vigorous mirror neuron system might lead to greater empathy, and thus, enhanced affiliation—including love. The inverse could also be true, if—as seems likely—living together and thus interacting on a regular basis provides opportunities for any existing mirror neuron system to be activated and/or further elaborated. Neuroplasticity research has already shown that the more neurons are stimulated, the better they function, and the more they grow: nerve cells that fire together, wire together. So maybe monogamy and mirror neurons are mutually predisposed. Never mind walking briefly—and metaphorically—in another's shoes: what about going alongside that person for literally decades? The more people stay together, their neurons firing in response to each other's experiences, the more likely they may be to empathize with, understand, and, yes, love each other.

Here is our final potential monogamy-promoting mechanism, or rather, molecule (actually, pair of molecules). In Shakespeare's *A Midsummer Night's Dream*, a love potion makes Titania, queen of the fairies, fall in love with a commoner (who, for added absurdity, had earlier been transformed into a donkey). Much hilarity ensues. A few centuries later, Wagner's *Tristan und Isolde* provides a more somber view of chemical cocktails and their powerful influence—on the human imagination if not directly on the libido. Indeed, the fantasy of an irresistible erotic brew has a history as ancient as it is compelling: Cupid's arrows were dipped in it, and anyone recalling popular music from the early 1960s will likely remember *Love Potion Number Nine*.

The truth, in this case, might be if not stranger than fiction, accurately depicted by it. There could really be a naturally occurring love potion. If so, it is our fourth Horseman. Start with two species of voles (small meadow mice of the genus *Microtus*). Prairie voles are among those few monogamous mammals in which male and female bond together, forswearing other sexual partners, and in which the male vigorously guards "his" female. Their close relatives, the montane voles, are more traditionally mammalian: interested only in one-night stands. Enter oxytocin, which, among other things, is a prosocial hormone.[44]

When natural oxytocin release is blocked, mammalian mothers reject their offspring. On the other hand, once infused with oxytocin, even virgin female rats fawn over the offspring of others—something that doesn't normally happen. Female prairie voles (the monogamous species) require prior contact with a male to be sexually receptive. In one experiment, virgin females were pretreated with oxytocin before being exposed to a male; they were sexually receptive right away, unlike controls, who received neutral saline and who rejected males' initial sexual advances.[45]

Furthermore, oxytocin doesn't only prime female prairie voles for sex, it is also key to their monogamous bonding. In fact, copulation isn't even necessary: introduce oxytocin into the brain of a female prairie vole, even if she hasn't mated, and the previously celibate vole will proceed to act like Titania or Isolde, enthusiastically bonded to the nearest male (although probably not if he's truly an ass). Moreover, if after copulation—when prairie voles typically establish their monogamous inclinations—the normal release of oxytocin is blocked, so is bonding.

The plot thickens. Oxytocin is also released in most mammals during labor and delivery; it helps generate uterine muscle contractions as the baby makes its way down the birth canal. Indeed, the name oxytocin derives from the Greek for "rapid birth," and a synthetic form of oxytocin known as pitocin is often administered to pregnant women to speed labor. Oxytocin is also involved in stimulating the "let-down" reflex when a nursing mother's breasts physically release their milk. Interestingly, Chinese midwives have long known of a connection between nursing and childbirth because tradition calls for applying ice to the nipples of women whose labor has stalled. Biologists now understand that stimulating the nipples leads to natural oxytocin release (an effect that, incidentally, is not only more natural, but also less out of control than what is produced by artificial pitocin "drips").

It probably isn't coincidental that the same naturally occurring hormone, oxytocin, has been recruited by evolution to facilitate both childbirth and milk release. Moreover, pair-bonding is an important ingredient here as well. It seems increasingly likely, in fact, that pair-bonding in many species—including humans—involves variations on a deeper theme: the mechanisms initially deployed to achieve mother–infant bonding. And here is a case in which *Homo sapiens*—despite its capacity for thought (or to some extent, because of it)—has special needs beyond that of other species.

I've noted, perhaps too often, that human beings are unusual in the helplessness of their infants. Accordingly, we don't just benefit from paternal involvement but are utterly dependent on new mothers bonding to their newborns, who, for example, and unlike other primates, cannot even cling effectively to their mother and therefore must be carried. Yet people are also unusual in the pain that women experience during childbirth, which likely confronted our ancestors with an evolutionary conundrum: how to induce the victim of so much hurt to affiliate with the source of the anguish?

Our species is blessed—sometimes, cursed—with a lot of intellect. We may be smart enough to acknowledge, rationally, the need for parental caretaking, but on the other hand, we are also intelligent enough to resent bestowing it on a tiny, odd-looking, squalling creature that has just caused so much pain. This dilemma has been solved by employing the same hormone, oxytocin, that generates uterine contractions to also induce lactation when the nipples are stimulated *and* to activate the same brain circuits that are involved in producing affiliation to another individual: namely, the source of that stimulus, the newborn infant.

The next step, and one that wouldn't have required much anatomic or biochemical innovation, is for this same hormone to be released during sexual intercourse in response to stimulation of the cervix and especially of the clitoris and leading to orgasm, as well as following erotic attention to female breasts, which is precisely what happens. And this, in turn, contributes mightily to bonding—now between mated adults.

Back to those voles, among whom oxytocin definitely is not the whole story. Unlike those quick-bonding prairie voles, montane vole adults don't bond, even when given artificial infusions of oxytocin. This is because female montane voles lack the brain receptors that are sensitive to this hormone. As for males, not surprisingly, their pair-bonding is less reliably linked to underlying biochemistry, if only because males don't have to get over the potentially profound bonding obstacle of childbirth. But their brain circuitry is similar to that of females, involving vasopressin, a hormone closely related to oxytocin. It turns out that pair-bonding (or lack thereof, in the case of the montane species) by male voles is predictable, depending on whether a particular genetic variant of a specific, identified vasopressin receptor gene is present. Montane voles don't have it, and they don't bond. Prairie voles do, and they do.

Give oxytocin *or* vasopressin to female or male montane voles and nothing happens—they lack the brain receptors that mediate response to the hormones. So it's a matter of whether the appropriate receptors are present: the hormones oxytocin and vasopressin occur in every mammal tested, including *Homo sapiens*. Presence or absence of receptors, on the other hand, is determined genetically, involving a gene that has been isolated and identified.[46] And this, in turn, likely correlates with socio-sexual bonding—monogamy—or its absence.

This is only part of the story, although as far as we know, it is the truth, the vole truth, and nothing but the truth. But it isn't limited to voles. Marmosets, which are monogamous, have more vasopressin chemically bound in their brains than do rhesus monkeys, which are promiscuous. And when the prairie vole vasopressin receptor gene was inserted into normally promiscuous mice, the recipients became more socially attached to their mates.[47]

But what about *Homo sapiens*?

The human species, interestingly, has the same gene for oxytocin and vasopressin receptors that is found in prairie voles and other mammals. Moreover, oxytocin and vasopressin aren't only released during childbirth and milk let-down, but also during sex. The actual attachment mechanism, in people no less than voles, appears to be connected to reward centers in the brain by which oxytocin receptors (in women) or their vasopressin counterparts (in men) stimulate the secretion of the neurotransmitter dopamine, which, in turn, feels good and induces voles, or people, to do more of whatever had generated the release in the first place. After a female prairie vole has mated, for example, researchers find a 50% increase in dopamine levels in her reward center.[48]

In addition, people who describe themselves as being madly in love show particular activation in the same brain regions that are stimulated by cocaine. So love can be a kind of addiction. As for oxytocin and vasopressin, they also seem to be especially involved in identifying specific individuals, and thus, connecting a rewarding bout of dopamine release with the rodent or person who helped evoke it. Among prairie voles, this recognition seems to involve olfactory cues in particular.[49] Moreover, oxytocin knockout mice (genetically engineered animals who lack the capacity to produce oxytocin) also lose the ability to recognize conspecifics, even though their general sense of smell remains unaffected.[50]

The role of oxytocin in identifying and bonding to specific individuals is not limited to prairie voles, nor to recognizing romantic partners; it extends to mother–infant attachment as well. Thus, domestic sheep "imprint" onto the odors of their newborn lambs, rejecting strangers. This, too, is achieved via oxytocin, which acts by neurotransmitter activity in the ewe's olfactory bulb, priming her brain to latch onto her baby's odor.[51]

Does something similar happen in people? Human beings are less smell driven than other mammals, but there is certainly much going on below the surface, unavailable to our conscious minds. And of course, there is every reason for sight, sound, touch, and a whole universe of other associations to combine and produce the end result, which often involves bonding—at least to some degree, and among some people.

All of which, in turn, leads to an important issue: variability. The simple fact that there are differences in the vasopressin receptor genes among voles leads to the suggestion that there is also variation in their sexual faithfulness. The inverse of the question—with or without an accompanying rock and roll beat—"Why do voles fall in love?" is bound to be "Why is it that some don't?" And, once again, what about people?

Sure enough, variability in marital and romantic relationships in our own species appears to correlate with the presence of different forms (technically known as "alleles") of the receptor gene. Men with one notorious genetic variant, for example,[52] remain single at twice the frequency of other men, and among those who do marry, are twice as likely to have had serious recent marital difficulties[53].

At this point, it should be clear that people have more than enough neural and hormonal infrastructure to support monogamy, evoked through evolution by the many biological advantages (including but not limited to co-parenting) that pair-bonding can provide. My guess is that the four mechanistic pillars described in this chapter fit together something like this. Human beings have a profound need for attachment, beginning in infancy and continuing through adulthood. The benefits of attachment, however, aren't limited to childhood, including as they do meaningful payoffs to adults as well. Attachment itself (at any age) is encouraged by standard psychological processes such as reward and reinforcement, and facilitated as well by mirror neurons, which, by promoting empathy, make for benevolent prosocial, interpersonal connections. All the while, these connections are being literally structured by the brain's capacity for neural plasticity in which nerve cells grow and brain regions develop in response to the continued interaction that defines attachment. And simultaneously, waiting in the wings, and ready to provide an encouraging chemical environment, are those love potion hormones, oxytocin and vasopressin, along with their gene-based receptors.

Much of the preceding is speculative, although plausible. It should at least be clear, however, that biology does not foreclose monogamy. Moreover, there are certain evolutionary factors—in addition to such proximate mechanisms as the Four Horsemen—that resonate strongly *with* monogamy, notably biparental (or at least, alloparental, multiadult) child care as well as various other efficiencies of scale and coordination. Two might not be literally able to live as cheaply as one, as the old saw had it, but like so many old saws, this one has retained some sharp teeth: two together can definitely live more cheaply than two, separately.

And then there is sex, not simply its hormonal consequences. Bachelors can have sex, and so can what used to be known as "spinsters." Compared to monogamists, both have, at least in theory, greater latitude when it comes to diversity—at least, diversity of partners. There is also no question that for its part, monogamy risks devolving into monotony. But there is also good evidence that married couples have more sex than do unmarried individuals, and even some indications that it tends to be better, too, especially for women. Thus, female orgasm is reported to be more reliable and more frequent with a known and trusted partner (whether that partner is male or female).[54]

Sex is certainly important, as a result of which "making love" can contribute to love being "made." But it is also important to distinguish between love and lust.

Here is novelist Milan Kundera's heterosexualized, male-oriented perspective on the whole complex business:

> Making love with a woman and sleeping with a woman are two separate passions, not merely different but opposite. Love does not make itself felt in the desire for copulation (a desire that extends to an infinite number of women) but in the desire for shared sleep (a desire limited to one woman).[55]

All that is needed to complete a tidy biological package is to add that after this shared sleep, one is likely to wake up—not only with one's sexual partner of the night before but also with a child or two thereby produced, and needing care.

The distinction between erotic lust and companionate love is often troublesome but is at least well-trodden in history, literature, and popular lore and understanding. By contrast, the coexistence of overt, identified, monogamous pair-bonding with covert, polygamous inclinations on the part of nearly every participant in these unions seems even more of an odd couple. But the human species is no stranger to odd coupling; one might even say that *Homo sapiens* is defined by such concatenations of opposites: divine and devilish, spiritual and material, good and bad, yin and yang, violent and pacific, emotional and intellectual, biological evolution and cultural teaching, and so on, ad infinitum.

Here, then, is something of a monogamy bottom line: human beings have several deeply evolved predispositions, which don't always coexist comfortably. In particular, we are almost certainly endowed with a strong inclination for pair-bonding, piled on top or alongside or incorporated within and sometimes without another, contradictory inclination: for multiple sex partners. To paraphrase Walt Whitman, do we contradict ourselves? Very well, we contradict ourselves. We are vast; we contain multitudes, which include simultaneous contradictory impulses toward both polygamy and monogamy.

Alexander Pope lamented our unique duality, noting in his *Essay on Man* (which applies no less to *Woman*) that

> He hangs between, in doubt to act, or rest,
> In doubt to deem himself a God, or beast;
> In doubt his mind or body to prefer;
> Born but to die, and reasoning but to err.

Pope concluded that we are creatures of preeminent paradox:

> Created half to rise and half to fall;
> Great lord of all things, yet a prey to all;
> Sole judge of truth, in endless error hurled:
> The glory, jest, and riddle of the world!

The push-me, pull-you contradiction between shared, publicly embraced, cultur-ally mandated dyadic monogamy on one hand and private, yet multipersonalized polygamy on the other isn't unique to the realm of mating partners. We live at the confluence of conflicting inclinations toward many things. Take violence, for exam-ple: human beings achieve remarkable feats of cooperation, communication, and conflict resolution and yet are also subject to aggressive, violent, even occasionally murderous responses under certain circumstances. The most appropriate symbol for humanity may well be the Roman god Janus, depicted with two faces—and who accordingly gave rise to the month of January, because one face looked back on the year that passed and the other forward to the one to come.

There is a story, believed to be of Cherokee origin, that captures much of this ambiva-lence, or perhaps ambidexterity, while also gesturing toward human agency as well as responsibility. A young girl is troubled by a recurrent dream in which two wolves fight viciously with each other. Seeking an explanation, she goes to her grandfather, highly regarded for his wisdom, who explains that there are two forces within each of us, struggling for supremacy, one embodying peace and the other, war. At this, the girl is even more distressed, and asks her grandfather who wins. His answer: "The one you feed."

Notes

1. Lukas, D., & Clutton-Brock, T. H. (2013). The evolution of social monogamy in mam-mals. *Science, 341*(6145), 526–530.
2. Reichard, U. (1995). Extra-pair copulations in a monogamous gibbon (*Hylobates lar*). *Ethology, 100*, 99–112.
3. Lukas and Clutton-Brock, "Evolution of Social Monogamy."
4. Opie, C., Atkinson, Q. D., Dunbar, R. I. M., & Shultz, S. (2013). Male infanticide leads to social monogamy in primates. *Proceedings of the National Academy of Sciences, 110*, 13328–13332.
5. Betzig, L. (1992). Roman polygyny. *Ethology and Sociobiology, 13*, 309–349.
6. Dr. Geertz was strongly opposed to biological interpretations of human behavior, a stance that caused us to cross swords on several memorable occasions.
7. This is reminiscent of the term "henotheism," coined by German philosopher Friedrich Schelling to describe Hindu religious practice: the worship of one god—at a time.
8. Lenski, G. E., & Lenski, J. (1978). *Human societies: An introduction to macrosociology* (3rd ed.). New York: McGraw-Hill.
9. Hrdy, S. B. (2011). *Mothers and others*. Cambridge, MA: Harvard University Press.
10. Jones, J. S., & Wynne-Edwards, K. E. (2000). Paternal hamsters mechanically assist the delivery, consume amniotic fluid and placenta, remove fetal membranes, and provide parental care during the birth process. *Hormones and Behavior, 37*, 116–125.
11. Wynne-Edwards, K. E. (1987). Evidence for obligate monogamy in the Djungarian hamster, *Phodopus campbelli*: Pup survival under different parenting conditions. *Behavioral Ecology and Sociobiology, 20*, 427–437.
12. Fernandez-Duque, E. (2007). The Aotinae: Social monogamy in the only nocturnal haplothines. In C. Campbell, A. Fuentes, K. C. MacKinnon, M. Panger, & S. K. Bearder (Eds.), *Primates in perspective*. New York: Oxford University Press.

13. Barash, D. P., & Lipton, J. E. (2002). *The myth of monogamy: Fidelity and infidelity in animals and people*. New York: Henry Holt.
14. Verner, J. (1964). Evolution of polygyny in the long-billed marsh wren. *Evolution, 18*, 252–261.
15. Alatalo, R. V., Carlson, A., Lundberg, A., & Ulfstrand, S. (1981). The conflict between male polygamy and female monogamy: The case of the pied flycatcher *Ficedula hypoleuca*. *American Naturalist, 117*, 738–753.
16. Smuts, B. (1985). *Sex and friendship in baboons*. New York: Aldine.
17. van Schaik, C. P., & Dunbar, R. I. M. (1990). The evolution of monogamy in large primates: A new hypothesis and some crucial tests. *Behaviour, 115*, 30–62.
18. de Waal, F. B. M. (2003). Apes from Venus: Bonobos and human social evolution. In F. B. M. de Waal (Ed.), *Tree of origin*. Cambridge, MA: Harvard University Press.
19. Fisher, H. E. (1989). Evolution of human serial pair-bonding. *American Journal of Physical Anthropology, 78*, 331–354.
20. Low, B. S. (2003). Ecological and social complexities in human monogamy. In B. Reichard & C. Boesch (Eds.), *Monogamy: Mating strategies and partnerships in birds, humans, and other mammals*. Cambridge, England: Cambridge University Press.
21. Marlowe, F. W. (2003). The mating system of foragers in the standard cross-cultural sample. *Cross-Cultural Research, 37*, 282–306.
22. Engels, F. (2010). *The origin and history of the family, private property and the state*. New York: Penguin Classics.
23. Knodel, J., Low, B., Saengtienchai, C., & Lucas, R. (1997). An evolutionary perspective on Thai sexual attitudes and behavior. *Journal of Sex Research, 34*, 292–303.
24. Slagsvold, T., Amundsen, T., Dale, S., & Lampe, H. (1992). Female-female aggression explains polyterritoriality in male pied flycatchers. *Animal Behaviour, 43*, 397–407.
25. Sandell, M. I., & Smith, H. G. (1996). Already mated females constrain male mating success in the European starling. *Proceedings of the Royal Society of London. Series B: Biological Sciences, 263*(1371), 743–747.
26. Alexander, R. D. (1987). *The biology of moral systems*. New York: Aldine de Gruyter.
27. Another one of those obvious "Duh!" facts that says a lot about human nature is that whereas most societies readily accept the need to have a license—to demonstrate basic competence—to drive a car, no such requirement exists for the much more demanding activity of rearing a child.
28. Chapais, B. (2008). *Primeval kinship: How pair-bonding gave birth to human society*. Cambridge, MA: Harvard University Press.
29. Hudson, V., & den Boer, A. M. (2005). *Bare branches: The security implications of Asia's surplus male population*. Cambridge, MA: MIT Press.
30. Keats' use of "awful" was intended to imply "generating awe," not something truly bad!
31. Jankowiak, W. (Ed.). (2008). *Intimacies: Love and sex across cultures*. New York: Columbia University Press.
32. Doctoral dissertation by H. Harris, cited in Pillsworth, E. G., & Haselton, M. G. (2006). Women's sexual strategies: The evolution of long-term bonds and extrapair sex. *Annual Review of Sex Research, 17*, 59–100.
33. Jankowiak, W., & Fischer, E. F. (1992). A cross-cultural perspective on romantic love. *Ethnology, 31*, 149–166.
34. De Rougemont, D. (1940). *Love in the western world*. New York: Pantheon.
35. Fisher, H. (2004). *Why we love: The nature and chemistry of romantic love*. New York: Henry Holt.

36. Hatfield, E., & Rapson, R. L. (2005). *Love and sex: Cross-cultural perspectives*. Lanham, MD: University Press of America.

37. Barash, D. P., & Lipton, J. E. (2009). *Strange bedfellows: The surprising connection between sex, monogamy and evolution*. New York: Bellevue Literary Press.

38. Every behavior requires an immediate "how" for it to occur; it requires an evolutionary "why" for the how to have been selected in the first place.

39. Bowlby, J. 1988. *A secure base*. Basic Books: New York.

40. Gaser, C., & Schlaug, G. (2003). Brain structures differ between musicians and non-musicians. *Journal of Neuroscience, 23*, 9240–9245.

41. Iacoboni, M., Woods, R. P., Brass, M., Bekkering, H., Mazziotta, J. C., & Rizzolatti, G. (1999). Cortical mechanisms of human imitation. *Science, 286*, 2526–2528.

42. Hickok, G. (2009). Eight problems for the mirror neuron theory of action understanding in monkeys and humans. *Journal of Cognitive Neuroscience, 21*, 1229–1243.

43. Dunbar, R. I. M., & Shultz, S. (2007). Evolution in the social brain. *Science, 317*, 1344–1347.

44. Carter, C. S. (1998). Neuroendocrine perspectives on social attachment and love. *Psychoneuroendocrinology, 23*, 779–818.

45. Cushing, B. S., & Carter, C. S. (1999). Prior exposure to oxytocin mimics the effects of social contact and facilitates sexual behavior in females. *Journal of Neuroendocrinology, 11*, 765–769.

46. Goodson, J. L., & Bass, A. H. (2001). Social behavior functions and related anatomical characteristics of vasotocin/vasopressin systems in vertebrates. *Brain Research Reviews, 35*, 246–265.

47. Young, L. J., Nilsen, R., Waymire, K. G., MacGregor, G. R., & Insel, T. R. (1999). Increased affiliative response to vasopressin in mice expressing the V1a receptor from a monogamous vole. *Nature, 400*, 766–768.

48. Young, L. J., & Wang, Z. (2004). The neurobiology of pair bonding. *Nature Neuroscience, 7*, 1048–1054.

49. Young, L., & Wang, Z. (2004). The neurobiology of pair bonding. *Nature Neuroscience, 7*, 1048–1054.

50. Donaldson, Z. R., & Young, L. J. (2008). Oxytocin, vasopressin, and the neurogenetics of sociality. *Science, 322*, 900–904.

51. Broad, K. D., Curley, J. P., & Kaverne, E. B. (2006). Infant bonding and the evolution of mammalian social relationships. *Philosophical Transactions of the Royal Society B: Biological Sciences, 361*, 2199–2214.

52. It is the 5′ flanking region of AVPR1A, allele RS3 334—if you must know.

53. Walum, H., Westberg, L., Henningsson, S., Neiderhiser, J. M., Reiss, D., Igl, W., . . . Lichtenstein, P. (2008). Genetic variation in the vasopressin receptor 1a gene (*AVPR1A*) associates with pair-bonding behavior in humans. *Proceedings of the National Academy of Sciences, 105*, 14153–14156.

54. Komisaruk, B. R., Beyer-Flores, C., & Whipple, B. (2006). *The science of orgasm*. Baltimore, MD: Johns Hopkins University Press.

55. Milan Kundera, *The Unbearable Lightness of Being*. New York: Harper Perennial, reprint edition.

6

Adultery

Infants have their infancy, and adults? Adultery. Even though monogamy is mandated throughout the Western world, infidelity is universal. Revelations of marital infidelity occur regularly, often among the most prominent individuals—most of them men—who have the most to lose: Bill Clinton, Newt Gingrich, Jesse Jackson, Mark Sanford, Elliot Spitzer, Tiger Woods. The list is enormous and is "updated" almost daily. In this chapter, I will examine the somewhat divergent motivations of men and women when it comes to infidelity—pointing out that in both cases, the underlying causes derive from internal whisperings of fitness maximization, underpinned by the fundamental biology expressed in our polygamous heritage.

In short, when adultery happens—and it happens quite often—what's going on is that people are behaving as polygynists (if men) or polyandrists (if women), in a culturally defined context of ostensible monogamy. Adultery, infidelity, or "cheating" are only meaningful given a relationship that is otherwise supposed to be monogamous. A polygynously married man—in any of the numerous cultures that permit such an arrangement—wasn't an adulterer when he had sex with more than one of his wives. (As candidate Barack Obama explained in a somewhat different context, "That was the point.") By the same token, a polyandrously married Tre-ba woman from Tibet isn't an adulteress when she has sex with her multiple husbands. Another way of looking at this: when people of either gender act on their polygamous inclinations while living in a monogamous tradition, they are being unfaithful to their sociocultural commitment, but not to their biology.

"Variability," wrote Christian apologist G. K. Chesterton,[1] "is one of the virtues of a woman. It avoids the crude requirement of polygamy. So long as you have one good wife you are sure to have a spiritual harem."[2] The problem—for Chesterton and others—is that many (perhaps most) men want a secular one.

Earlier, when describing the basic biology of male–female differences, we considered the Coolidge Effect. There is a large body of literature commenting on it, and on the tendency for men in particular to equate monogamy with monotony. Lord Byron wondered "how the devil is it that fresh features/Have such a charm for us poor human creatures?" Speaking more delicately, W. S. Gilbert, in *Trial by Jury*, alluded to the flip side of the male fondness for variety with the knowing line, "Love unchanged will cloy." Three hundred years earlier, Shakespeare had described Cleopatra as follows: "Age cannot wither her, nor custom stale her infinite variety."

But then, Cleopatra was supposed to have been remarkable precisely because by contrast, "other women cloy the appetites they feed."

Once more, the ultimate mechanism of all this "cloying" is likely to be found in the adaptive advantage gained by turning some proportion of male sexual energy toward new exploits and thus, more potential evolutionary success. As to its proximate mechanism, we can only guess. At the level of brain cells and neurochemicals, we do know that repeated stimulation can result in a degree of insensitivity—the flip side of attachment, discussed in chapter 5. It is therefore possible that something happens with respect to sexual enthusiasm analogous to the habituation that occurs with, say, the constant hum of a refrigerator motor: after a time, people habituate to the noise and only notice it when it stops! Thus, perhaps after a prolonged sexual association (perhaps weeks, months, even years) brain cells—and male brain cells in particular—might simply become habituated: that is, saturated with neurotransmitters, or refractory to them.

Here is yet another—and, I announce with regret, the last—extended quotation from the estimable and curmudgeonly Mr. Mencken, this time elaborating on the boredom that can accompany monogamy, and how it might be assuaged:

> Monogamous marriage, by its very conditions . . . forces the two contracting parties into an intimacy that is too persistent and unmitigated; they are in contact at too many points, and too steadily. By and by all the mystery of the relation is gone, and they stand in the unsexed position of brother and sister. . . . A husband begins by kissing a pretty girl, his wife; it is pleasant to have her so handy and so willing. He ends by making Machiavellian efforts to avoid kissing the every day sharer of his meals, books, bath towels, pocketbook, relatives, ambitions, secrets, malaises and business: a proceeding about as romantic as having his boots blacked. The thing is too horribly dismal for words. Not all the native sentimentalism of man can overcome the distaste and boredom that get into it. Not all the histrionic capacity of woman can attach any appearance of gusto and spontaneity to it.
>
> [O]nce the adventurous descends to the habitual, it takes on an offensive and degrading character. The intimate approach, to give genuine joy, must be a concession, a feat of persuasion, a victory; once it loses that character it loses everything. Such a destructive conversion is effected by the average monogamous marriage. It breaks down all mystery and reserve, for how can mystery and reserve survive the use of the same hot water bag and a joint concern about butter and egg bills? What remains, at least on the husband's side, is esteem—the feeling one has for an amiable aunt. And confidence—the emotion evoked by a lawyer, a dentist or a fortune-teller. And habit—the thing which makes it possible to eat the same breakfast every day, and to windup one's watch regularly, and to earn a living.
>
> [One might] prevent this stodgy dephlogistication of marriage by interrupting its course—that is, by separating the parties now and then, so that neither will become too familiar and commonplace to the other. By

this means, . . . curiosity will be periodically revived, and there will be a chance for personality to expand a cappella, and so each reunion will have in it something of the surprise, the adventure and the virtuous satanry of the honeymoon. The husband will not come back to precisely the same wife that he parted from, and the wife will not welcome precisely the same husband. Even supposing them to have gone on substantially as if together, they will have gone on out of sight and hearing of each other. Thus each will find the other, to some extent at least, a stranger, and hence a bit challenging, and hence a bit charming. The scheme . . . has been tried often, and with success. It is, indeed, a familiar observation that the happiest couples are those who are occasionally separated, and the fact has been embalmed in the trite maxim that absence makes the heart grow fonder. Perhaps not actually fonder, but at any rate more tolerant, more curious, more eager. Two difficulties, however, stand in the way of the widespread adoption of the remedy. One lies in its costliness: the average couple cannot afford a double establishment, even temporarily. The other lies in the fact that it inevitably arouses the envy and ill-nature of those who cannot adopt it, and so causes a gabbling of scandal. The world invariably suspects the worst. Let man and wife separate to save their happiness from suffocation in the kitchen, the dining room and the connubial chamber, and it will immediately conclude that the corpse is already laid out in the drawing-room.

Obviously, a fondness for sexual variety can lead to adultery. Just as obviously, however, it doesn't have to. But as the famous team of sex researchers led by Dr. Alfred Kinsey pointed out

Most males can immediately understand why most males want extramarital coitus. Although many of them refrain from engaging in such activity because they consider it morally unacceptable or socially undesirable, even such abstinent individuals can usually understand that sexual variety, new situations, and new partners might provide satisfactions which are no longer found in coitus which has been confined for some period of years to a single sexual partner. . . . On the other hand, many females find it difficult to understand why any male who is happily married should want to have coitus with any female other than his wife.[3]

Recall the comical dialog between Anna and the King of Siam from chapter 3. This man–woman disparity is not simply because society has typically sought to repress female sexual desire (although it has, and for reasons that are also consistent with biology) but because most women do not experience heightened lust simply upon being presented with a new, anonymous partner. Again, at the ultimate, evolutionary level this is almost certainly because a new partner *as such* is unlikely to enhance a woman's reproductive success; and so, women have not been outfitted with a comparable "Mrs. Coolidge effect." Certainly, a woman is *capable* of sexual intercourse with new and different men—sometimes, as in the case of prostitutes, many different men in succession—but this is quite different from being inspired to do so by the very newness of the partner. Indeed, the fact of sexual variety is itself cited by prostitutes as one of the emotionally deadening aspects of their work.[4]

Nevertheless, men have not cornered the market on infidelity. If nothing else, for every adulterous man there must be at least one sexually willing woman—although this individual need not be married herself. We know that women's sexual inclinations are not simply the converse of men's, with utter monogamous fidelity the opposed mirror image of randy male infidelity. But we also know that because of the greater physical size, strength, and potential violence of men, women are inclined to be especially secretive when it comes to their philandering.

Biologists have long known that monogamy is rare in the animal world, especially among our fellow mammals. But we didn't have any idea how truly rare it is until DNA fingerprinting arrived on the scene and was applied to animals in the mid-1990s. Previously there had been hints, but these were largely ignored. For example, in an attempt during the 1970s to reduce excessive numbers of blackbirds without killing them, a substantial number of territorial males were surgically sterilized. To the surprise of the researchers, many female blackbirds—mated to these vasectomized males—produced perfectly normal offspring.[5] Evidently, there was hanky-panky going on within seemingly sedate blackbird society. Nonetheless, for decades it has been the received wisdom among ornithologists that 92% of bird species were monogamous.

Over time, however, a new realization dawned: social monogamy—in which a male and a female court, spend time together, and set up joint housekeeping—is not the same as sexual monogamy, that is, limiting copulation to one's social partner. Not that members of a socially monogamous pair don't have intercourse with each other, it's just that often they *also* do so with others. Hence the term Extra-Pair Copulations (EPCs) was born, now a standard concept in animal behavior research, as studies employing DNA fingerprinting have revealed, time and again, that even those species that seemed devotedly monogamous were only socially so, and sexually? Not so much. Depending on the species, it is common to find that from 10% to 60% of avian offspring are not fathered by the mother's social partner.[6]

Biologists were not surprised to find evidence that males are prone to sexual gallivanting. After all, even though the basic biology of sperm makers doesn't quite mandate searching for and when possible indulging in multiple sexual partners, it clearly inclines males of most species in that direction. What was perplexing, however, was the finding that females—even those in apparently stable domestic unions—were similarly disposed. It's just that they are more secretive about it. As a result, biologists (such as me) would spend hundreds of hours carefully watching the behavior of a mated pair of birds without seeing any indication of sexual infidelity by the female—not necessarily because she was sexually faithful to her mate but because she was hiding her EPCs: not from researchers, but from her social partner. Why was she doing this? It was for reasons not dissimilar from why socially monogamous human beings are comparably secretive when it comes to their own EPCs.

Evidence has accumulated from a variety of animal species that if and when the male finds evidence that "his" female has been associating with other males, he is prone to do the animal equivalent of refusing to provide child support: no longer provisioning or defending the offspring, presumably because they might not be genetically his. We have already considered male–male violence, especially in the context of a man encountering his wife's paramour (chapter 2). Although male–male violence is also found in animals, it is rare for nonhuman males to attack their mate upon discovering her infidelity. Most commonly, he punishes her—biologically, he defends his own fitness—by abandoning the "bastard" offspring, a response that can be devastating for the success of those offspring, and hence, for the "unfaithful spouse." This pattern has been found in mammals[7] as well as birds[8]. And although there do not seem to be any data dealing explicitly with the effect of suspected adultery on divorce and disputed child support in human beings, common sense suggests a close connection.

Not all organisms are equally prone to EPCs, if only because some species don't form pair bonds in the first place. Others, such as the flatworm parasite, *Diplozoon paradoxum*, found in freshwater fish, are strictly monogamous: male and female encounter each other while adolescents, after which their bodies literally fuse together, until death do they not part. (Hence the genus name, *Diplozoon*, indicating two animals, while *paradoxum* is self-explanatory.) Other animals are only sexually receptive for very brief intervals, greatly reducing the opportunity for sexual exploration; female giant pandas, for example, are only in estrus for two to three days per year. Human beings, on the other hand—women no less than men—are endowed with 24/7/365 sexual potential, providing immense opportunities: for fitness enhancement or diminution, exciting adventure as well as miserable heartbreak.

There are remarkably few reliable DNA data on the actual frequency of human extrapair paternity, despite the fact that such testing is now widely available. Maybe this paucity shouldn't be surprising because of possible reluctance on the part of most men to question their wife's fidelity. As a result, the available information tends to be biased toward circumstances in which there is liable to be higher than average extrapair paternity, such as divorce proceedings, when child support is in dispute. In any event, the frequency of human marriages in which the husband is not the genetic father of the mother's children range from as low as 0.03 to 11.8%.[9]

I t is one thing to understand why females—of pretty much all species—hide their EPCs from their social partners, especially if biparental cooperation is expected when it comes to rearing offspring. More mysterious is this: given the potential costs of being caught, why do females engage in any EPCs at all? Bear in mind that because eggs are produced in very small numbers, whereas sperm are astoundingly abundant, females very rarely need to copulate with more than one male to be fertilized. It turns out that biologists have identified a number of potential benefits accruing to a "cheating" female, with the specifics varying with the species and sometimes with individual circumstances.

Here are a few of the major fitness payoffs thus far recognized:

- Increase the genetic variety of their offspring
- Obtain more desirable (i.e., fitness-enhancing) genes for their offspring than can be provided by their social mate
- Obtain additional resources, notably food, from their "lovers"
- Enhance their social status by affiliating with a male who is more dominant than their current partner
- Purchase "infanticide insurance" by inducing other males to behave parentally toward the females' offspring
- Explore the potential of switching from a less desirable to a more desirable partner

One of the enduring mysteries of animal behavior and evolution has been why some species show considerable sexual dimorphism even though their primary mating system is socially monogamous. For example, a neotropical bird known as the resplendent quetzal is truly resplendent, so much so that it is the national bird of Guatemala, with its image on that country's flag and coat of arms. But only the male is endowed with resplendently shimmering feathers and a dramatically elongated tail; female quetzals are relatively drab. Now that even social monogamy is revealed to be rife with polygynous and polyandrous departures, we can speculate as to the basis of such dimorphism, which might well owe its existence to the fact that the male quetzal's resplendence, for example, likely enables him, at least on occasion, to achieve additional matings outside his seemingly monogamous union.

There is no reason why similar considerations (obtaining additional resources, better genes, etc.) couldn't motivate human females, too, although in the case of women, other factors could be involved as well—which typically are not assumed to operate among animals. It is worth noting that these causes are all "proximate," although they each have straightforward ultimate underpinnings. Moreover, although biologists have had less reason to seek explanations for EPCs by men than by women (because the biology of sperm making provides more than enough evolutionary rationale), the following considerations could apply to both sexes:

- Retaliating for infidelity by one's partner
- Responding to other sources of anger with one's partner
- Short- or long-term fascination or interest in a particular lover
- Seeking sexual or social gratification not otherwise available in the primary relationship

A bottom-line, take-home message is that when sexual infidelity occurs among human beings—and whether the "infidel" is a woman or man—it is because a fundamental, biologically generated, polygamous inclination (polygyny in the case of men, polyandry in the case of women) has broken through the existing monogamous social structure. Not surprisingly, men consistently report higher levels of sexual infidelity in marriage than do women. This, in turn, conforms to the biology of maleness versus femaleness discussed earlier, and with such findings as a study that encompassed 52 different countries and about 16,000 respondents, and found that men consistently expressed interest in having more sexual partners than did women.[10] But a male–female difference in adultery could also be due, at least in part, to the near-universal double standard in which men are socially encouraged to be more sexually adventurous and to seek multiple partners so as to be seen as "real men," whereas women who are acknowledged to be similarly inclined are often denigrated as "loose," "easy," or "sluts." Some men, as well, may be liable to exaggerate their number of infidelities, just as some women may be inclined to understate theirs.

It is pretty much a cross-cultural universal that men intimidate their spouses to refrain from extramarital sex, punishing them—often severely and not uncommonly, lethally—should they do so. Aside from their internal motivations, women may be most prone to "cheating" under the following circumstances:

1. When they have the support of their own relatives, which is especially likely in societies that are "matrilocal"—that is, when women live near their extended families—as opposed to "patrilocal" societies in which following marriage, the bride moves in with her husband's family. Most human societies are patrilocal, which means that a wife is surrounded by her husband's kin. This makes it easier for a man to keep tabs on her, and itself facilitates a double standard. It is quite possible that patrilocality became the most common marital living arrangement precisely because it serves to discourage a wife's infidelity, thereby reassuring husbands.
2. When wives don't rely very heavily on their husbands for material support, protection, and assistance in childrearing. Cross-cultural studies have found that among societies in which women get most of their sustenance from relatives, rather than from their husbands, they are more likely to have extramarital affairs and are also more prone to divorce.[11]
3. When their husband is less "desirable" than others who are sexually available, with "desirability" assessed either biologically or socioeconomically. Interestingly, something comparable occurs in at least one bird species. Research on EPCs in black-capped chickadees found that females were more likely to "cheat" with males who are socially dominant, and especially when their current mate is relatively subordinate. Where females have the opportunity to learn the relative rank of all neighboring males with respect to her own mate, females regularly pursue the strategy of seeking out EPCs with superior partners. Although lower-ranked males may suffer temporary losses through the EPCs of their mates, each male has some chance of attaining alpha rank if he lives long enough. Once at alpha rank, a male will likely engage in more EPCs himself, while having a mate that will no longer seek EPCs elsewhere.[12]

Not just chickadees cheat: a study of sexual behavior in modern China found that women whose husbands' income was lower than the median were more likely to engage in extramarital sex.[13]

Even in societies that explicitly permit extramarital sex (and there are a few), infidelity is nearly always carefully circumscribed and is not simply permitted willy-nilly. A now-classic review of human sexuality from a cross-cultural perspective concluded that

> With few exceptions . . . every society that approves extra-mateship liaisons specifies and delimits them in one way or another. There are some peoples, for example, who generally forbid extra-mateship liaisons except in the case of siblings-in-law. This is true among the Siriono, where a man may have liaisons with his mate's sisters and with his brother's wives and their sisters. Similarly, a woman has sexual access to her husband's brothers and the husbands of her sisters. . . . In some societies extra-mateship liaisons take the form of "wife lending" or wife exchange. Generally, the situation is one in which a man is granted sexual access to the mate of another man only on special occasions. . . . Another type of permission in respect to extra-mateship liaisons appears in some societies in the form of ceremonial or festive license . . . [ranging] from harvest festivals to mortuary feasts.[14]

Human sexual practices are notably diverse, but mostly with regard to rules about which partners are suitable, permissible, desirable, recommended, or prohibited, sometimes including details as to frequency of coitus and potential physical positions. But hopes, or—in some cases—fears that the primordial human condition is one of bonobo-like sexual promiscuity are simply not justified by anything ever theorized by biologists or found by anthropologists. (More about this in chapter 9.)

On the other hand, there is a growing body of experimental evidence to suggest that women partake of a "dual-mating strategy," consisting of both long-term and short-term tactics. The former involves establishing a bonded relationship with a consistent partner, typically someone able to invest sufficiently in offspring and also predisposed to do so, whereas the latter calls for responding positively—especially when ovulating—to men who are literally perceived as "sexy," which is to say, possessing good genes (discussed initially in chapter 3). Evidence for this dual strategy comes largely from a diversity of studies showing that when they are most fertile, women are especially predisposed to prefer images, sounds, even smells arising from men who are higher in testosterone, who possess greater body symmetry—in short, who are likely to offer "good genes."[15]

At the same time, it must be emphasized that many of these findings involve laboratory assessments, and their highly artificial conditions may or may not reflect what people really do. Technical questions about the details of these studies go beyond the scope of this book; suffice it to say that they appear to be effectively resolved. To put it baldly and admittedly with some oversimplification, there appears to be at least a faint female predisposition to marry the more androgynous, "good father" type but dally with the bad boy stud.[16]

It is nonetheless possible that human physiology gives a positive payoff to couples remaining together, not just long enough to produce a child, but to have achieved a level of physical intimacy along the way, which would predispose against the short-term mating strategy just described. Preeclampsia, a form of hypertension, results from immunological disparity between mother and fetus; it can be a serious complication of pregnancy and occurs roughly 10% of the time. The risk of preeclampsia decreases with increased duration of a woman's sexual relationship with a given partner, evidently because the woman's immune system becomes increasingly habituated to the seminal products of a given man and therefore less liable to a potentially dangerous immune response when she is carrying an embryo that contains 50% of his genes.[17] If this scenario is valid, then mating with a new, short-term partner and promptly producing a child would increase the risk of this complication. This, in turn, would tend to mitigate against the adaptiveness of a "dual mating strategy."

As already described, among some Amazonian peoples in particular, polyandry is facilitated by a belief in "partible paternity," the biologically inaccurate but superficially logical notion that a child can have multiple fathers, consisting of the men who had intercourse with the mother during her pregnancy.[18] It is probably significant that the majority of these societies are matrilocal (husbands reside with the wife's relatives), so the women have social support. This is important because even when paternity is thought to be partible, human sexual jealousy is such that the woman's designated husband is typically unenthusiastic about sharing his paternity as well as his wife.

When it comes to the Seven Deadly Sins, cavorting along with anger, greed, sloth, pride, lust, and gluttony, we find "envy" (in Latin, *invidia*). Not "jealousy," although in fact jealousy is a whole lot more deadly than envy. Jealousy is also a whole lot more biologically motivated. Jealousy and envy are close, but not identical. A useful distinction is that "Envy concerns what you would like to have but don't possess, whereas jealousy concerns what you have and do not want to lose."[19] You might be envious of someone who has a rich, attractive spouse, but jealous if your own partner seems to be interested in her or him. You could envy the person who "has" this spouse, but at the same time, you don't want to lose the partner you currently have, and would be jealous if he or she were unfaithful.

An evolutionary perspective shows that this anxiety about possible loss is, at balance, worry about losing fitness, something that is particularly acute if it involves loss of an otherwise reproductively enhancing relationship. It doesn't matter, by the way, if you and your partner have firmly decided not to have children, or if secure birth control measures were followed during any EPCs; just as people are to some degree biologically inclined toward polygamy—for themselves—they are equally predisposed *against* similar inclinations by their partners. Our biology operates largely independent of our cognitive intent, just as women ovulate and men produce sperm, whether or not they intend to become parents.

Sinful or not, sexual jealousy is certainly real, and is particularly evoked in the aftermath of real or imagined adultery. It is also found in women no less than men, although for perfectly "good" biological reason, the male version tends to be more cross-culturally prevalent as well as more violent. In his recent book, *Jealousy*, classicist Peter Toohey[20] has unearthed a number of ancient imprecations reflecting male jealousy, including this one from second-century Egypt in which a betrayed husband begs the gods to "let burning heat consume the sexual parts of [his wife], including her vulva, her members until she leaves the household."

We can be quite confident that *Homo sapiens* did not evolve in a social environment like that of chimpanzees or bonobos in which lots of sperm competition took place. For one thing, our testis size isn't anything like that of chimps or bonobos. In addition, as described in chapter 3, the anatomical structure of human sperm argues strongly against our species having a multimale, multifemale sexual heritage. On the other hand, however, we are provided with substantial amounts of sexual jealousy, not just in the Judeo-Christian West, but cross-culturally, a behavioral adaptation that presumably wouldn't exist if it weren't called for.

The Ten Commandments are clear when it comes to not coveting your neighbor's wife; and although it is intriguing that no comparable warning proscribes coveting your neighbor's husband, there is little doubt that sexual covetousness—by either sex—is more risky than the simple material kind. Coveting your neighbor's lawn mower may be bad, but in the annals of covetousness and its consequences, it could be worse. In the Draconian precepts of Islamic *sharia* law, an adulterer can readily lose his or her life, whereas a thief will lose only his hand.

There are some—albeit rare—human societies in which married women are granted social permission to engage in extramarital sex, mostly with a sibling of their spouse. However, I don't know of any human groups in which women are granted more sexual freedom than are men. For much of human history, adultery was defined by a pronounced double standard: sexual relations between a *married woman* and a man other than her husband. Such cases have been—and still are—widely seen as offenses *against the woman's husband*. By contrast, if a husband has sexual relations with an unattached woman or with a prostitute, the majority of cultures do not consider this adultery . . . so long as the woman in question isn't married to another man.

There is a genetic factor—actually, an array of them—that predispose toward marital infidelity in human beings. A version of the dopamine receptor gene (DRD4), occurs on chromosome 11 and is found in all people, although individuals vary in how many times this gene is repeated: from 2 to 11. People with 7 or more repeats of DRD4 turn out to be represented in greater proportion than would be expected due to chance alone among those engaging in extramarital sex.[21] However, this isn't an "infidelity gene" but rather a genetic predisposition toward greater sensation seeking. One might expect that individuals with multiple repeats of DRD4 would also be more likely to go skydiving, or to enjoy roller coasters. Nor is it that they necessarily have a higher sex drive, or a genetic proclivity to extramarital sexual exploits as such; rather, they crave novelty.

It is one thing to insist on something new for dinner every night, quite another to insist on a new lover. One way of conceptualizing the problem—without explicitly invoking biology—is that such behavior violates what in the Western tradition is known as "social contract theory." The idea is fundamental to much political philosophy, including the work of Hugo Grotius, Thomas Hobbes, and especially John Locke, whose *Second Treatise of Government* (1698) laid out the proposition that people agree to live convivially in a social unit—as large as a nation-state or as small as a domestic family—by forgoing certain individualistic options to gain other benefits, based on cooperation and shared responsibilities. Just as government "derives its just powers from the consent of the governed" (a basic principle employed a century and a half later by the framers of the US Constitution), marriages derive their legitimacy and stability from the consent of the participants. And one of the most important such consensual necessities is sexual fidelity. The husband–wife sociosexual contract is thus the governmental social contract writ small.

Under its terms, and taking a hard-eyed look at its contractual aspects, women provide men with a guarantee of their sexual fidelity as well as a partner for regular intercourse along with other shared domestic payoffs, while men provide women with resources, protection, and assistance in childrearing. Along with a mutual sharing of genes. Although this traditional contract was—and still is—unfair in its implied asymmetry with regard to sexual fidelity, it was in many ways an excellent one. There is an additional problem, however, which is the substance of this book: both men and women carry with them an evolutionarily generated inclination to violate the contract and to consort with other partners—that is, for polyandry as well as polygyny. Biologists have if anything been late to the party when it comes to appreciating the potential of both polygyny and polyandry to assert themselves despite a sociocultural commitment to monogamy. This recognition is part of a new and important realization on the part of evolutionary biologists: men and women often have distinctly different evolutionary interests. This is true despite the fact that human beings are if anything unusual among living things in that their interests are likely to be shared when it comes to caring for their needy offspring. We are stuck with, on one hand, a biological basis for a biparental social contract, and, on the other, a no less biological basis for polygamous yearnings.

It takes two, not just to tango but to engage in adultery. And often, the two aren't equally motivated. Accordingly, there occurs what evolutionary psychologists call "mate poaching." One study identifying the various tactics used by would-be mate poachers found that more than one-half the couples interviewed had at one time attempted to "steal" another's spouse, either for a short- or long-term relationship.[22]

What about reactions to adultery? Both sexes—at least in the United States—react strongly and negatively, men somewhat more than women. Of women, 64% employ some form of physical violence; 88% of men do so—and more violently.[23] Here is poet Carl Sandburg, in a brief yet complete poem that he called a "novel":

> Papa loved mama
> Mama loved men
> Mama's in the graveyard
> Papa's in the pen

The opera *Carmen* gives us a dramatized version of this same scenario, one that has long resonated in the human imagination. In Bizet's ever-popular masterpiece,

Don Jose, a man with a "short fuse," is fascinated by a beautiful, powerful woman. A battle ensues, the central question being control over the woman herself, which is to say, over her sexual behavior. Carmen—sultry, seductive, and free spirited—would rather die than be dominated. In the end, she does just that: rather than lose sexual control of Carmen, Don Jose kills her.

Why is there so much truth within the story of Carmen, or Sandburg's grim humor? Why do men so often react violently when confronted with an adulterous wife? Biology provides a cogent cross-cultural answer, even though real-life participants as well as authors of great fiction don't necessarily understand the evolutionary precepts under which they are operating. Much of the double standard is straight ("unadulterated") biology, combined with social rules that have historically looked after the interest of—you guessed it—men. The fact is that women get pregnant; men don't. "The difference is boundless," said Samuel Johnson (1709–1784): "The man imposes no bastards upon his wife." An adulterous man may impose pregnancy on his lover, but so long as the lover is not married, no *man* is shortchanged—that is, fooled into rearing children that are not biologically his—as a result. But if an adulterous wife becomes pregnant, then her husband may unknowingly rear another man's offspring. ("Mommy's babies, Daddy's maybes," once again.)

The derivation of the word adultery is itself revealing: it comes from the Latin *adulterare*, meaning alter or change. To adulterate means to "debase by adding inferior materials or elements; making impure by admixture." The crucial admixture in this case is another male's genes. Although this is understandably troublesome from the males' viewpoint, females have a different perspective. A recent review[24] of 260 mammal species concluded that the male reproductive strategy of infanticide (itself characteristic of polygynous species subject to periodic male takeovers) is to some extent countered by the female strategy of multiple sexual partners. In species in which females mate with many different males, these males develop unusually large testes. By virtue of what the researchers call "paternity dilution," females essentially induce males to compete with their sperm rather than by killing each other's offspring—because it isn't entirely clear whose infants are whose, and just as selection favors disposing of unrelated offspring, it inhibits destroying one's own.

The word "cuckold" is also revealing, once again as a statement of male sexual and parental insecurity. It comes from the European cuckoo, renowned for its behavior as a "nest parasite." Female cuckoos lay their eggs in the nest of other species, who then become unwitting hosts. When the cuckoo chicks hatch, they add injury to insult by ejecting the host's biological offspring, thus monopolizing its foster parent's resources for itself. For a man to be cuckolded is to suffer the fate of those unwitting male birds who fail to see they have been displaced by a lover, and end up not only biological failures but social laughingstocks.

In *Love's Labour Lost*, Shakespeare gives us this: "The cuckoo then in every tree; Mocks married men; for thus sings he, "Cuckoo; cuckoo; cuckoo; O word of fear; unpleasing to a married ear!" When a man's parental investment is expended on behalf of another man's child, his love's labor is lost indeed.

Paternity dilution as a female strategy for preventing male-initiated infanticide can thus be a two-edged sword: although men who have had sexual relations with a woman may indeed—like their animal counterparts—be less likely to be subsequently infanticidal toward her offspring, such a strategy risks generating not only abandonment but outright aggression from a man who suspects that he has been cuckolded.

Even during the French Revolution, at a time when enthusiasm for creating a new society was so great that the names for months of the year were tossed out and replaced with new ones, sexual asymmetries of the old regime were retained in one regard: legal sanctions against a *wife's* adultery. And in modern times men are far more likely than women to cite adultery as a cause for divorce. In a sample of 104 societies, historian Laura Betzig found that a wife's infidelity was a primary cause of divorce in 48; male infidelity, in not a single one.[25]

In Tolstoy's great novel, *Anna Karenina*, a liberal-minded gentleman named Pestov comments that the real inequality between husband and wife is the fact that infidelity by each is punished differently. Karenin responds, "I think the foundations of this attitude are rooted in the very nature of things." In fact, asymmetric punishment for adultery hardly exhausts the catalog of husband–wife social inequality. Nonetheless, both Pestov and Karenin are correct: adultery by a wife is indeed punished differently than is adultery by a husband.

Nowhere on earth do husbands take a light, carefree view of their wives' adultery, except perhaps in fiction. In Aldous Huxley's *Brave New World*, "everyone belongs to everyone else," and sexual jealousy is abolished, as shown by this hymn to promiscuity:

Orgy-porgy, Ford and fun,
Kiss the girls and make them One.
Boys at one with girls at peace;
Orgy-porgy gives release.

It may give release, but genuine sexual promiscuity does not seem to bring peace. The easygoing sexuality of Huxley's *Brave New World* makes for interesting speculation, but nothing like it has ever been found, at least nothing that has lasted for any appreciable time, because it would require a brave—and essentially nonbiological—new human being as well. It is noteworthy that in *Brave New World*, it is "John the Savage" who retains a biologically vibrant insistence that sex should be associated with emotional intensity. Margaret Mead's accounts of Samoa—long held out as an example of natural "free love"—have been debunked.[26] The United States witnessed its own "free love" movement in the 1960s, but like all such movements, it was short-lived and generally fraught with emotional tension and instability; when the flower children of the 1960s grew up, most of them married and became rather conventional serial monogamists. Adultery, of course, persisted, but "free love" turned out to be emotionally and often financially expensive.

The more wives a man has, the more he must concern himself with the fidelity of his wives, as evidenced by the eunuchs whose function was to guard a sultan's harem, or the court ladies of ancient China who were tasked to devise elaborate techniques for keeping track of the Emperor's many wives and concubines. Another technique was foot binding, which served a double purpose; it demonstrated that the family was wealthy enough to do without the work of the foot-bound mistress of the house, and also kept her housebound, making it difficult for her to get away, for nonmarital sex or anything else. Other societies—especially in Afghanistan, India, Pakistan, and Saudi Arabia—have kept women in "purdah," or strict seclusion, effectively isolating them from other men. In northern India, high-status women (Hindus as well as Muslims) have long been cloistered, imprisoned in their own homes.

One of the most notorious and excessive offshoots of this whole nasty business is the brutal and—by Western standards—grotesque practice of female circumcision, a ritual form of genital mutilation that makes Judeo-Islamic male circumcision (removal of the foreskin) look like fingernail clipping. Female circumcision is a potentially life-threatening and often debilitating form of genital mutilation. Tens of millions of young girls—especially in East and West Africa—have been subjected to various forms of this procedure, which typically involves removal of the clitoris (clitoridectomy) and sometimes closing up of the vagina (infibulation), either by sewing its lips together or by severe scarring, caused by deliberately cutting or burning the vaginal wall. An opening is left for menstruation, but one too small to admit a penis. Upon marriage, an infibulated woman is cut open to permit sexual relations with her husband. Sometimes she is even restitched if the husband must be away for an extended time.

Although these behaviors are interwoven with much cultural tradition, there can be little question that their ultimate function is essentially biological, with men as ever seeking to control the sexuality of women: removing the clitoris greatly reduces a woman's sexual pleasure, while infibulation serves as a grotesquely effective, built-in premarital chastity belt. Women who refuse these procedures are liable to be socially scorned and considered unmarriageable. Although it is older women who nearly always enforce the rules and actually carry out these procedures, they are acting out what is fundamentally a demand of men, one to which they themselves were typically subjected in their own childhood. Ironically, given the social rules by which they operate, the genital mutilation of their daughters may well be in the parents' best reproductive interest, secured by making their daughters marriageable, which in turn is thought to require proof against any likely adultery.

In 18th-century Geneva, under the sway of John Calvin, premarital sex on the part of an engaged couple was a serious offence, warranting legal punishment. But this was nothing compared with "premarital adultery," sex with a third party; this was intolerable, doubly so if the bride-to-be had been a virgin. It was a capital crime. In one of his sermons, Calvin explained why this was so heinous:

> The woman desecrated her soul and body. She stole her future husband's exclusive rights to her, which he had purchased from her father. She heaped on her future husband the obligation to raise her illegitimate child. She deprived the couple's own later legitimate children of precious family resources. And

such a fundamental breach of faith so early in the relationship would inevitably lead to similar serious sins. Both her husband and the broader society would be better off if this adulterous fiancée were executed.[27]

Interestingly, Calvin also demanded that "The male fornicator should be punished equally," although it isn't clear that this was ever the case. Male fornication (along with adultery, so long as the woman involved wasn't another man's wife) has always been seen as a different matter, something akin to "boys will be boys" and the "sowing of wild oats." Here is a work by John Suckling (1609–1642), whose self-mocking poem "The Constant Lover" made fun of his variety-seeking libido while also expressing it:

> OUT upon it, I have loved
> Three whole days together!
> And am like to love three more,
> If it prove fair weather.
>
> Time shall moult away his wings
> Ere he shall discover
> In the whole wide world again
> Such a constant lover.
>
> But the spite on 't is, no praise
> Is due at all to me:
> Love with me had made no stays,
> Had it any been but she.
>
> Had it any been but she,
> And that very face,
> There had been at least ere this
> A dozen dozen in her place.

It is difficult to imagine anything like Suckling's jovial self-indulgence, were a woman to celebrate, with a comparable wink-and-a-nod, her own diverse love affairs. And this distinction is, if anything, enhanced when it comes to EPCs, aka adultery, infidelity, or "cheating." According to Kinsey and colleagues, the double standard is both widespread and ancient: "wives, at every social level, more often accept the non-marital activities of their husbands. Husbands are much less inclined to accept the non-marital activities of their wives. It has been so since the dawn of history."[28]

Just as male aggression and sexual behavior are closely linked, so is male aggression and sexual jealousy. The Old Testament speaks repeatedly of a "jealous God," from which, even if we knew nothing else, we can assume it to be male, a god intolerant of sharing worship, of "having other gods," not just *before* this one, but alongside or behind, or anywhere else at all. In just this way, husbands are particularly disinclined to allow their wives to have other men, either before them, after them, on the side, or any other way . . . while at the same time, as we have seen, they often seek precisely such liaisons for themselves.

Is this double standard a cultural artifact? Certainly, it is amplified by social traditions and expectations; it doesn't emerge automatically from the demands of our DNA, like Athena emerging fully armed from the head of Zeus or analogous, for example, to intestinal contractions after eating. But as with male–female differences in aggressiveness, the double standard is found in every culture where it has been looked for, and even among animals. It is worth repeating that historically and around the globe, adultery is viewed as a crime against *men*, that is, against the husband whose wife is adulterous; and so it is toward her male lover that much of the husband's anger is directed, although nearly always, wives, too, are punished when they are adulterous. The fate of the male adulterer—the man who has sex with another man's wife—is often castration or death. In contrast, when it is the woman who is wronged, that is, when her husband commits adultery, there are few, if any, actual penalties, *provided he is not "wronging" another married man in the process.*

There is simply no comparison between what befalls a married woman who has sex with an unmarried man and a married man who engages comparably with an unmarried woman (if caught, that is). Indeed, such attitudes extend back to the Egyptians, Hebrews, Babylonians, Romans, Spartans, and so forth, who defined adultery strictly by the marital status of the *woman*. If no man is "wronged," then essentially no wrong is supposed to have been done.

In which mating system—monogamy or polygamy—is sexual infidelity more frequent? The evidence is not persuasive, and a theoretical case can be made either way. Under monogamy, each female is associated with a male, and therefore, philandering by either member of such a pair, although potentially beneficial to the fitness of the philanderer, is detrimental to the one left behind. Hence, we'd expect that under monogamy, extrapair matings might be comparatively rare because the two "partners" are likely to be closely watching each other. In fact, one way of maintaining sexual fidelity within monogamous marriage is for the husband and wife to agree to a kind of "mutually assured destruction" treaty (call it "mutually assured monogamy"), analogous to Cold War deterrence: "I won't have an affair and drive you crazy, in return for which you promise not to have an affair and drive me crazy." Otherwise, the marriage could explode.

Alternatively, we might expect that polygyny would diminish philandering, if only because the males in question are physically and or culturally more imposing (recall that sexual dimorphism is characteristic of polygyny in general). Moreover, don't forget that gorillas, large and bodily intimidating, have very small testicles because their highly competitive bodies make it unnecessary for them to compete via their sperm. On the other hand, it is presumably more difficult to keep track of multiple females, unless the harem holder has access to large numbers of committed underlings, who would themselves need to be desexed if they are to be trusted.

It turns out that there is no obvious correlation between polygyny and the frequency of extrapair copulations, at least in birds.[29] And in human beings? Your guess is as good as mine.

Female sexual agency is now and has always been greater than one might think; and although women are often victimized in the course of polygyny, their sex has a long history of assertiveness. As Mel Konner—evolutionary anthropologist and physician—writes

> Female primates were never passive in any situation, . . . [including] resisting male abuses. They were constantly active on their own and their kids' behalf, jockeying for position in a female dominance order, competing over males, food and territory, shielding their young from harm and trying to ensure that they would grow up into roles where (male or female) they would be dominant, too. If some primate females had to live in a male world—and not all did—they would not just stand by; they would make the most of it.[30]

Among chimpanzees, high-status females will sometimes mate with several males, including lower-ranked individuals when the females are less fertile, which may be an attempt to confuse paternity or to secure favors such as grooming. Even so, knowing the alpha male's reputation for violence, the deed is almost always done quickly and on the sly, away from the eyes of the dominant male. And when high-status females are ovulating, they show a strong preference for mating with dominant males.[31]

Unapologetic, in-your-face polyandry among human beings is pretty much limited to a bimodal sample: women at the upper end of the socioeconomic distribution who flaunt their sexuality and get away with it (such as the singer Madonna, Lady Gaga, or Elizabeth Taylor who had seven husbands) or those at the other end, notably prostitutes whose vocation requires what looks like polyandry but is more like a job—or slavery. Even though advances in contraception have enabled women to free themselves from the reproductive tyranny that had previously restricted their sexuality, it is safe to say that the great majority of women are less liberated than basic fairness would prescribe.

Nonetheless, change is, ironically, the most reliable human constant, if not in our biology, then in its cultural accompaniments. In February 2015, South Korea's Constitutional Court—citing the country's changing sexual mores and a growing emphasis on individual rights—struck down a law that for more than six decades had made adultery a criminal offense, punishable by up to two years in prison. Share prices for a leading Korean condom manufacturer, *Unidus*, immediately rose by nearly 15%; and *Hyundai Pharmaceutical*, which markets so-called morning-after birth control pills, rose in value by nearly 10%.

Notes

1. Chesterton, G. K. (1911). *Alarms and discursions*. New York: Dodd, Mead and Co.
2. This oft-quoted quip actually appeared as a throwaway aside in a delightful essay in which Chesterton was commenting on the appealing variability of a woodlot visible from his home:

There is a line of woodland beyond a corner of my garden which is literally different on every one of the three hundred and sixty-five days. Sometimes it seems as near as a hedge, and sometimes as far as a faint and fiery evening cloud. The same principle (by the way) applies to the difficult problem of wives.

3. Kinsey, A., Martin, C., & Pomeroy, W. (1948). *Sexual behavior in the human male*. Philadelphia: W. B. Saunders.
4. Moran, R. (2015). *Paid for: My journey through prostitution*. New York: W. W. Norton & Co.
5. Barash, D. P., & Lipton, J. E. (2002). *The myth of monogamy: Fidelity and infidelity in animals and people*. New York: Henry Holt.
6. Ibid.
7. Richardson, P. R. K., & Coetzee, M. (1988). Mate desertion in response to female promiscuity in the socially monogamous aardwolf, *Proteles cristatus*. *South African Journal of Zoology, 23*, 306–308.
8. Cezilly, F., & Nager, R. G. (1995). Comparative evidence for a positive association between divorce and extra-pair paternity in birds. *Proceedings of the Royal Society of London Series B: Biological Sciences, 262*(1363), 7–12.
9. Simmons, L. W., Firman, R. C., Rhodes, G., & Peters, M. (2004). Human sperm competition: Testis size, sperm production and rates of extrapair copulations. *Animal Behaviour, 68*, 297–302.
10. Schmitt, D. (2003). Universal sex differences in the desire for sexual variety: Tests from 52 nations 6 continents, and 13 islands. *Journal of Personality and Social Psychology, 85*, 85–104.
11. Gaulin, S., & Schlegel, A. (1980). Paternal confidence and paternal investment: A cross-cultural test of a sociobiological hypothesis. *Ethology and Sociobiology, 1*, 301–309; Flinn, M. V., & Low, B. S. (1986). Resource distribution, social competition, and mating patterns in human societies. In D. Rubenstein & R. Wrangham (Eds.), *Ecological aspects of social behavior*. New York: Plenum Press.
12. Smith, S. M. (1988). Extra-pair copulations in black-capped chickadees: The role of the female. *Behaviour, 107*, 15–23.
13. Zhang, N., Parish, W. L., Huang, U. Y., & Pan, S. (2012). Sexual infidelity in China: Prevalence and gender-specific correlates. Archives of Sexual Behavior, 41, 861–873.
14. Ford, C. S., & Beach, F. A. (1951). *Patterns of sexual behavior*. New York: Ace Books.
15. Gangestad, S. W., Thornhill, R., & Garver-Apgar, C. E. (2005). Adaptations to ovulation. In D. M. Buss (Ed.), *The handbook of evolutionary psychology*. Hoboken, NJ: Wiley.
16. Gildersleeve, K., DeBruine, L., Haselton, M. G., Frederick, D. A., Penton-Voak, I. S., Jones, B. C., & Perrett, D. I. (2013). Shifts in women's mate preferences across the ovulatory cycle: A critique of Harris (2011) and Harris (2012). *Sex Roles, 69*, 516–524.
17. Robillard, P. Y., Dekker, G. A., & Hulsey, T. C. (2002). Evolutionary adaptations to pre-eclampsia/eclampsia in humans: Low fecundability rate, loss of oestrus, prohibitions of incest and systematic polyandry. *American Journal of Reproductive Immunology, 47*, 104–111.
18. Walker, R. S., Flinn, M. V., & Hill, K. R. (2010). Evolutionary history of partible paternity in lowland South America. *Proceedings of the National Academy of Sciences, 107*, 19195–19200.
19. von Sommers, P. (1988). *Jealousy: What it is and who feels it?* New York: Penguin.
20. Toohey, P. (2014). *Jealousy*. New Haven, CT: Yale University Press.

21. Garcia, J. R., MacKillop, J., Aller, E. L., Merriwether, A., Wilson, D. S., & Lum, J. K. (2010). Associations between dopamine D4 receptor gene variation with both infidelity and sexual promiscuity. *PloS ONE, 5,* e14162.

22. Schmitt, D. P., & Buss, D. M. (2001). Humane mate poaching: Tactics and temptations for infiltrating existing mateships. *Journal of Personal and Social Psychology, 80,* 894–917.

23. Jankowiak, W., & Hardgrave, M. D. (2007). Individual and societal responses to sexual betrayal: A view from around the world. *Electronic Journal of Human Sexuality, 10.*

24. Lukas, D., & Huchard, E. (2014). The evolution of infanticide by males in mammalian societies. *Science, 346*(6211), 841–844.

25. Betzig, L. (1989). Causes of conjugal dissolution: A cross-cultural study. *Current Anthropology, 30,* 654–676.

26. Freeman, D. (1983). *Margaret Mead and Samoa: The making and unmaking of an anthropological myth.* Cambridge, MA: Harvard University Press.

27. Witte, J., & Kindgdon, R. (2005). *Sex, marriage, and family in John Calvin's Geneva.* New York: Wm. B. Eerdmans.

28. Kinsey, A. C., Pomeroy, W. B., & Martin, C. E. (1948). *Sexual behavior in the human male.* Philadelphia: W. B. Saunders.

29. Møller, A. P. (2003). The evolution of monogamy: Mating relationships, parental care and sexual selection. In U. H. Reichard & C. Boesch (Eds.), *Monogamy: Mating strategies and partnerships in birds, humans, and other mammals.* Cambridge, England: Cambridge University Press.

30. Konner, M. (2015). *Women after all.* New York: W. W. Norton.

31. Wroblewski, E. E., Murray, C. M., Keele, B. F., Schumacher-Stankey, J. C., Hahn, B. H., & Pusey, A. E. (2009). Male dominance rank and reproductive success in chimpanzees, *Pan troglodytes schweinfurthii. Animal Behaviour, 77,* 873–885.

7

Genius, Homosexuality, and God:
Three Hypotheses

It might be true—and I emphasize *might*—that polygamy has tendrils that reach farther than human mating systems alone. In this chapter, we will consider three hypotheses that speculate about possible connections between polygamy and (1) genius, (2) homosexuality, and (3) God. Let's start with the troublesome question of genius, and particularly its evident male bias.

At one point in her essay, *A Room of One's Own,* Virginia Woolf imagined the life of a hypothetical Judith Shakespeare: "She was as adventurous, as imaginative, as agog to see the world as he was. But she was not sent to school." Instead, Judith was relegated to cooking and cleaning, and consigned to an unwanted marriage wherein her highest (indeed, pretty much her only) acceptable aspiration was to bear children. She eventually killed herself without ever writing a word. What we might call the Woolfian Hypothesis is so obviously true that it hardly needs elaborating: women have traditionally been suppressed by patriarchy, and whatever genius they possess has not been permitted anything like the flowering that men's genius has enjoyed.

This is perhaps the most cogent explanation for the following vexatious generalization. Among recognized geniuses, there have been immensely more men than women. Given that we know beyond a doubt that women are every bit as intelligent and creative as men, this disparity, and the mere possibility that biological differences may in any way be responsible, emerges as an awkward conundrum that most people—the liberal minded especially—would prefer to ignore.

Oscar Wilde pointed to a seeming paradox in the gendering of genius. "Owing to their imperfect education," he wrote, "the only works we have had from women are works of genius."[1] It is often unadvisable to "expand" on the works of Mr. Wilde (himself no slacker in the genius department), but in this case it might be worth clarifying his intent, which was to emphasize that given the cultural oppression under which women have labored, it may well have required works not just of genius but of surpassing, unassailable genius for their contributions to be made, not to mention noticed.

It is an embarrassment to liberal, feminist sensibilities, and yet, the historical record is undeniable: when it comes to the ranks of identified geniuses, men outnumber women by a huge margin. Where are the women to rank with Cezanne, Monet, Raphael, Rembrandt, Turner, Vermeer? Sculptors whose works come anywhere near those of Bernini, Brancusi, Donatello, Michelangelo, Rodin? Playwrights of the order of Chekhov, Euripides, Ibsen, Moliere, Racine, Shakespeare (William)? Composers whose works compare to Bach, Beethoven, Brahms, Mozart, Schubert, Stravinsky, or Tchaikovsky? Poets in the same ballpark as Byron, Coleridge, Eliot, Pushkin, Whitman, Wordsworth, Yeats? Scientists of the stature of Archimedes, Copernicus, Darwin, Einstein, Galileo, Newton? Mathematicians whose insights can stand comparison to those of Euler, Fermat, Gauss, Hilbert, Riemann? Novelists who stand with Balzac, Cervantes, Dickens, Dostoyevsky, Faulkner, Joyce, or Tolstoy? Foundational religious figures of the status of Confucius, Jesus, Lao-tse, Mohammed, Moses, Siddhartha Gautama, Zoroaster? Where are the Renaissance women to rank with the Renaissance men such as Leonardo or Michelangelo? Of course, it can be an entertaining—and remarkably challenging— parlor game to come up with renowned women in each of these creative arenas (the most readily identified probably coming from the ranks of poets and novelists, at least in modern times), but the pattern is clear: there are precious few "old mistresses" to compare with the embarrassing preponderance of "old masters."

The Woolfian Hypothesis aside, there are several imaginable explanations for this dramatic disparity. For example, perhaps the mere identification of "genius" versus "normal"—never mind the difference in sex ratio among purported geniuses—simply constitutes a distinction without a difference, a Western individualistic construct that doesn't say anything at all about human universals. Unfortunately, it is currently impossible to evaluate this possibility. I, at least, have been unable to unearth any data that speak to the ratio of identified male versus female geniuses cross-culturally; and so, my information at present derives almost entirely from the Western experience.

Another possibility is that maybe the difference—even if it turns out to be a genu- ine pan-human phenomenon—is more apparent than real: perhaps there have been innumerable Marie Curies, Hildegards of Bingen, and Frida Kahlos. It's just that when genius is displayed by women, it isn't taken as seriously as when the creators happen to be carrying a Y chromosome. "We cannot know if there is such a thing as altogether unappreciated genius," wrote Hannah Arendt in an essay dealing with literary reputations, "or whether it is the daydream of those who are not geniuses." I'll go farther: we cannot know for sure if there even is such a thing as genius . . . full stop. One person's genius is often another's crackpot. Freud comes immediately to mind.

There are many works of the imagination—literary, musical, visual, even scientific— that are today considered masterpieces and the products of genius but were ignored or disparaged in their day. One thinks, for example, of Melville's *Moby Dick*, Stravinsky's *The Rite of Spring*, the paintings of van Gogh, poems by Keats, stories by Kafka, the theories of Ludwig Boltzmann (which gave rise to statistical thermodynamics and even- tually, quantum mechanics). Genius itself is a fraught concept, and maybe one that is simply more likely to be applied to men than to women, regardless of their underlying creative capacities, or even their outward accomplishments. This is certainly a hypothe- sis worth entertaining, although I suspect that it won't withstand serious scrutiny. Think of Justice Potter Stewart's famous observation about pornography: I may not be able to define it, but I know it when I see it. Given the opportunity, most of us see genius, with what appears to be great clarity. (It must be emphasized at the same time that most of us have very little clarity when it comes to those aspects of social bias—italicized by Woolf's imagined Judith Shakespeare—that impinge on the visibility of genius.)

Add the concept of "muse." Nearly always, a genius's muse is considered to be female, virtually never male. For Dante it was Beatrice; for John Stuart Mill, Harriet Taylor; for Yeats, Maude Gonne; for Vladimir Nabokov, Vera; for John Lennon, Yoko Ono. In some cases—for example, photographer Alfred Steiglitz and painter Georgia O'Keefe—the "muse-ness" was roughly symmetric and mutual, but nearly always, a muse has been female and the person thereby inspired male. A man can be a "boy toy," just as a woman can be a "trophy wife"; but even though few people invoke a muse these days, a male muse is pretty much an oxymoron.

Maybe—an unlikely hypothesis, but worth mentioning nonetheless—this whole genius business is itself bogus. Maybe everyone is equally talented, with luck and connections driving certain individuals to achieve renown: that is, everybody is a winner, and everyone deserves a trophy. Welcome to Lake Woebegone, where all the children are above average, but only a fortunate few actually get rewarded. Then there is the least plausible notion of all, the rigidly deterministic, ridiculously misogynistic perspective from blinkered biology, which holds that there must be something in the Y chromosome—for all its depauperate DNA—or in testosterone (found, incidentally, in women as well as men, albeit usually in smaller quantities), or in both, that predisposes at least some men to be exceptionally talented, while the rest of us (male and female) are left to stumble on as merely average.

To this diverse farrago let's add another possible explanation, one that is admittedly speculative and that doesn't necessarily exclude any of the others but which is also fully consistent with evolutionary biology, while at the same time respecting the unquestionable role of cultural expectations and restrictions so poignantly captured in Virginia Woolf's tale of Judith Shakespeare. Thus, the Woolfian Hypothesis is almost certainly the strongest of those described here, doubtless accounting for a substantial proportion of the male–female divergence. But given the stubborn persistence of this divergence, both historically and geographically, it seems worthwhile to consider whether it derives at least partly from something in addition to culture; namely, our biology—although not in the crudely deterministic way that most readily comes to mind.

Allowing ourselves to think in terms of the human evolutionary bequeathal has the advantage that it subsumes the major difficulty with the Woolfian Hypothesis: if woman geniuses are no less abundant than their male counterparts but have been squelched by arbitrary cultural rules and social traditions, there should be as many societies in which men are similarly inhibited and in which female genius flourishes. If, as its advocates typically claim, certain male–female distinctions are purely a consequence of arbitrary cultural rules, roles, and traditions, then, because they are arbitrary, they should be randomly distributed across different societies. On the other hand, insofar as the male–female "genius disparity" appears to be a cross-cultural universal, this would seem, in turn, to call for an equally cross-cultural (i.e., biological) contribution.

It could be argued that cultural processes are indeed arbitrary (and thus, in a sense, nonbiological) yet also universal, if they are due to some other cross-cultural universal, a male–female difference that underlies and drives the resulting pan-human homogeneity. But any such universal male–female difference must be "biological" as well. Our shared biology as members of the species *Homo sapiens* is precisely what all human beings, regardless of their cultural trappings, have in common.[2]

The following hypothesis attempts to reconcile the apparent reality of male prominence when it comes to identified genius with the self-evident role of culture, giving full credence to the intellectual equality of men and women. It revolves around the default setting of human polygyny and one of its less obvious consequences. We might call it *A Portrait of the Genius as a Show-Off*.

The gendering of genius would begin, not surprisingly, with the fact that *Homo sapiens* is evolutionarily predisposed to polygyny. It is unlikely that male elk, gorillas, or elephant seals qualify as "geniuses," or that they are more genius prone than are females of their species. Compared to human beings, most animals are limited in the dimensions by which a would-be harem master can distinguish himself.

But distinguish themselves they do. When they aren't butting heads or otherwise competing mano-a-mano, among males of most polygynous species (which is to say, most males of most mammal species), courtship consists of various ways to show themselves off. The classic example is the male peacock, among which it is worth noting that males are equipped not only with elaborately attention-grabbing tail feathers, but with a behavioral disposition to strut and dance so that these secondary sexual characteristics can be emphasized, thereby impressing not only other peacocks but also the peahens. Here is an account of the peacock's display:

> A peacock struts his stuff slowly, arcing great turquoise plumes that dwarf his glistening blue body, raising a patch of iridescent gold coins, then sweeping a delicate green mesh up into a lustrous fan dotted by gorgeous, staring green-and-gold eyes, in which the bird stands onstage alone, radiating a gaudy spray with feathers like the sun's rays, only in color. Another turn or two later, he enfolds himself in drapery, collapsing his sumptuous feathers down into a sleek, pied multicolored tail that seems to loll along behind him endlessly.[3]

Other animals do other things, but the crucial point for our purposes is that those cavorters, attention-seekers, singers, strutters, chest beaters, and otherwise exhibitionistic self-displaying creatures are nearly always male.

Evolutionary biologists aren't certain why females find such antics appealing (after all, there are liabilities to being so conspicuous and to investing so much metabolic energy in the process), although it is clear that they do. One likely explanation, known as the "sexy son hypothesis," described in detail in chapter 3, suggests that females are attracted to show-offy males because by mating with them, these females are likely to produce sons who will, upon maturity, be comparable show-offs and thus sexually attractive to the next generation of females, thereby providing an eventual fitness payoff to the initial, star-struck females. Breeding with show-off males produces show-off sons who, because they are sexually appealing, generate more grandchildren for the original chooser. This, in turn, would make for selection pressure on the males to be both physically structured and behaviorally inclined to draw attention to themselves.

Among *Homo sapiens,* males aren't outfitted with anatomical traits comparable to the peacock's tail, the lion's mane, or the silverback gorilla's massive chest or sagittal crest. But—according to my hypothesis—men are predisposed to distinguish themselves from their competitors, not so much by their anatomy as by their behavior, which includes their creativity. Unfortunately once again, the relevant data simply aren't available, but anecdotal anthropological reports strongly suggest that cross-culturally, the sex inclined to show off is overwhelmingly male. They are the ones who engage in lengthy song contests (Inuit and Australian aborigines), competitive gift giving (Northwest Coast native Americans), display of hunting trophies (numerous people, from Africa to urbanized America), and so forth.

When it comes to showing off, it seems only logical that one's mental products would be included. After all, the brain may well be—at least in part—the human equivalent of the peacock's tail, an organ that could well have been overdeveloped because through evolutionary time, it's possessors were the beneficiaries of sexual selection. To make a long story short (one especially well explained by Geoffrey Miller, in his stimulating and original book *The Mating Mind*[4]), many of the accomplishments of the human brain, such as complex language, poetry, music, painting, philosophy, sculpture, dance, theoretical science, and so forth, have no obvious survival value but make sense as the outcome of mate choice whereby those possessing sufficient brain power were perceived by our great-great-grandmothers as being downright sexy. One way of looking at this, in terms known to students of animal behavior as the "handicap principle," is that you have to be a pretty impressive dude to be able to invest in all that otherwise useless brainpower!

Accordingly, men have invented numerous ways to show off how impressive they are, to literally separate ourselves from the crowd. Chief among them are numerous creations of the mind, which leads to the additional intriguing possibility that many of our notable accomplishments—especially those that aren't obviously functional—may have been generated by the adaptive payoff associated with distinguishing those who produce them. I submit that the minds of women are every bit as capable of such creativity as are the minds of men, but that to an extent not commonly found in women, men are encouraged, or in some cases, actually driven by their biology to act on, and, moreover, to publicize, whatever distinctive qualities they possess. And of course, such behavior is encouraged—even admired—in many cultural traditions as well, in ways that counterpart activities by women are not. When a man seeks to get ahead and to call positive attention to himself, he is often lauded as "ambitious," whereas when a woman behaves similarly, she is liable to be derided as a "bitch."

In short, regardless of their inherent qualities, men are generally more inclined than women to distinguish themselves from the crowd of other males, or at least, to try. Moreover, because of polygyny—in which a disproportionate number of males are exceptionally successful whereas the rest aren't—they are more likely to be assertive, sometimes downright aggressive, and stubbornly persistent about whatever gives them a claim to greater distinctiveness. Even the iconic solitary genius, who appears to pursue his passion simply for its own sake, and who isn't consciously seeking recognition, could nonetheless be in the throes of acting out an evolved predisposition, one that was biologically rewarding to his ancestors.

Australian biologist Rob Brooks[5] has pointed out that even in societies in which both men and women contribute food to the family unit, there is a gender disparity that can best be understood in the context of the biology of male–female differences. Brooks points out, for example, that among the inhabitants of Mer Island, between northern Australia and Papua New Guinea, both men and women fish. Whereas women concentrate on foraging for shellfish as well as catching in-shore fishes, men concentrate on the larger pelagic species. This difference isn't surprising given that men have somewhat more upper body strength. However, as Brooks emphasizes, because of their fishing techniques, women actually bring back more total calories than do men; in other words, if men chose to do so, they could be considerably more effective as providers.

As Brooks writes

> Whereas most of the fish caught by Merian women go directly to feeding their families, the big spectacular fish that men occasionally catch are more likely to be part of the tradition of elaborate feasts, which . . . involve dancing, distribution and display. This difference is typical of many hunting societies in which the food that women gather or catch directly feeds their families whereas men divide the meat from any large animal they have killed among all the households.

A key point here is that such dividing and sharing by men isn't nearly as benevolently altruistic as might be thought. Rather, it serves a reproductively selfish function: making themselves more attractive to women (as well as perhaps raising their status among other men), which in turn often leads to more mating opportunities:

> Men often hunt the largest and most elusive animals—large pelagic fish among the Merian, bison in pre-colonial North America, armadillos and peccaries in South America, kudu, zebra and eland in Africa, and possibly even mammoths in ice-age Europe. By chasing such big prey, men actually bring in far fewer calories than they would have done by targeting smaller, easier to catch and less dangerous animals. Instead of aiming for hunting efficiency, men target large animals because by killing such a beast and sharing the meat they burnish the legend of their hunting prowess, and this leads to better mating opportunities in the future. An extra mating can far outweigh, in evolutionary terms, the mundane fitness advantage of providing a few thousand calories to one's own wife and children.

Noting that our knowledge of prehistoric livelihoods is due in large part to the spectacular cave paintings of Chauvet and Lascaux, Brooks points out that whereas these examples of prehistoric art focus heavily on large animals and various hunting attempts, there are essentially no depictions of successful gathering. "It strikes me," he writes, "that many cave paintings may have been attempts to keep fresh the memories of the most epic hunts long after the meat had disappeared."

A similar pattern has been found among many current traditional societies—and may well be characteristic of most. Thus, among the Aché, men hunt for peccary and search for honey from wild bees (both dangerous and unreliable activities, but with high potential payoff), while women gather fruit, insect larvae, and pound starch from palm nuts. The payoff to a successful hunter can be a genuine bonanza—40,000 calories or more if he bags a peccary—but on average, an Aché hunter brings home 9,634 calories per day, with his median return being a paltry 4,663 calories. By contrast, an Aché woman contributes 10,356 calories.[6] So why don't men forego hunting for the more reliable returns that women achieve?

Kristen Hawkes, the anthropologist who documented these male–female differences, discusses them in terms of "provider" versus "show-off" strategies. Women are the providers; men, the show-offs. A similar bifurcation occurs among the Hadza of east Africa: men go off hunting giraffes while women stay home preparing mongongo nuts, a highly reliable and nutritious protein source, but one that lacks social panache. And among both the Aché and Hadza (and also, interestingly, among chimpanzees who succeed in killing red colobus monkeys or juvenile baboons[7]), even when a hunt is successful, the returning hero does not act as a reliable provider for his own family; rather, the meat obtained by a successful hunter is typically shared with other tribe members, thereby increasing the "generous" hunter's likelihood of obtaining additional wives or, at minimum, an extra copulation or two. In short, he burnishes his polygyny.

The benefits of simply obtaining calories are real enough, but as engineers might put it, they don't "scale up." Rather, they are subject to diminishing returns, so that getting 20 times more than the usual amount of calories from an especially successful hunt does not, in itself, result in 20 times more reproductive success: except if it helps you get 20 wives! (or "just" 20 additional copulations, especially if they are with different women).

Men, like males generally, aren't only drawing attention to themselves because women are watching and responding with applause that also sometimes includes sexual access; they may also be contributing to their status among other men. Male strategy in most polygynous systems emphasizes vertical organization along with doing whatever it takes to obtain power and then to keep it. Here is primatologist Frans De Waal, describing the situation among chimpanzees: "The unreliable, Machiavellian nature of male power games implies that every friend is a potential foe, and vice versa. Males have good reason to restore disturbed relations; no male ever knows when he may need his strongest rival."[8] Accordingly, successful show-offs would not only gain by attracting females and thus, mating opportunities, but in addition, they would be more likely to attract opportunistic coalition partners among other males.

Females, on the other hand, are more likely to inhabit a world of horizontal relationships based especially on kinship connections and reciprocal altruism, oriented mainly toward greater success for their offspring. Ambrose Bierce notoriously defined politics as "the conduct of public affairs for private advantage." Even today, in an era of increasing gender equality, such conduct is largely a guy thing, with males primarily concerned with personal status-seeking and females with benefitting their friends and genetic relatives.

According to de Waal, differences in social goals between male and female chimps

> May explain most of the discussed sex differences. In other words, both sexes may have the same mental abilities and social tactics available but do not use them in the same way because the kind of outcomes they try to achieve are different. This might mean that the resemblance between human and chimpanzee coalition patterns is due to similar social goals in the lives of women and female chimpanzees, on the one hand, and in the lives of men and male chimpanzees, on the other. If so, the two species would share a personality difference between the two sexes that is very profound indeed.[9]

In short, females and males have somewhat different goals when it comes to their politicking, with females focusing more on social stability and mutually supportive relationships in a smoothly functioning network and males being more concerned with their personal rank within the social system. If the accomplishments of genius are motivated—consciously or not—by attempts at self-aggrandizement, then the apparent male bias when it comes to "genius" emerges as a by-product of polygyny and *not* as a result of any inherent gendering of genius itself.

Intriguing support for this comes from a small-scale but fascinating "experiment" conducted in the late 1960s by a Hungarian school teacher named László Polgár.[10] He had written a book titled *Bring Up Genius!* in which he argued that with appropriate training and suitable experiences, any child could be exceptionally successful in any field. He and his wife, Klara, chose competitive chess as an arena in which to test their idea, using their own children as subjects. There had never been a woman grandmaster, and it was widely assumed—by male chess experts and commentators—that women somehow lacked the intellectual rigor and/or suitably combative inclinations to compete at the highest levels.

The Polgárs had three daughters to whom they taught the basics of chess. The results were stunning. Their eldest daughter, Zsuzsa, became in 1991 the first woman grandmaster in history. Their middle child, Zsófia, became an International Master; and—wait for it!—their youngest, Judit, became at age 15 the youngest grandmaster (of either sex) and eventually achieved victories over Magnus Carlsen, Garry Kasparov, and Boris Spassky. It seems unlikely that the Polgár girls owed their accomplishments to genetics alone, because their mother, Klara, didn't play at all, and their father, László, was at best a mediocre player. The choice of chess as the arena in which the Polgár girls were to excel was purely arbitrary.

In any event, it is absolutely clear that Judit Polgár and her sisters were not kept from success by their additional X chromosomes. The preponderance of male over female chess "geniuses" is almost certainly due to the cultural environment and social encouragement given to boys over girls when it comes to this particular endeavor.

And quite possibly, it is also due to the social as well as biological predisposition on the part of successful male chess players—along with their friends, family, and teachers—to excel and to trumpet their accomplishments.

By the way, the hypothesis that male genius is largely (and perhaps entirely) due to a social bias ultimately linked to a polygynous past is not refuted by the fact that some iconic male geniuses—for example, Leonardo, Beethoven, Newton—were childless. Others were prolific reproductively no less than creatively: Einstein had three children, Picasso had four by three different women (not to mention his numerous lovers), and Bach had no fewer than 20. Moreover, it makes perfect sense that natural selection would have endowed men with a proclivity to act on whatever talents they possess by being internally rewarded for performing such actions, independent of whether they were also rewarded romantically and/or reproductively.

There is, finally, another interesting insight generated by the hypothesis that the bias toward male genius may result from an underlying, biologically facilitated tendency for men in particular to show off whatever "natural talent" they happen to possess or are capable of nourishing, just short of sprouting peacock-like plumage. It revisits a variant of the Woolfian Hypothesis, harking back as well to the notion whereby genius is—whatever else it may be—also socially constructed. Thus, could it be that genius itself is a societal notion, created primarily by men, developed *to reward showing off?* If so, then among the various "inherent" traits with which geniuses have been endowed, and for which they are rewarded, being a successful show-off would be high on the list; therefore, geniuses would tend—among other things—to be particularly successful show-offs, regardless of the objective quality of whatever they are preening, promoting, and self-publishing about. Perhaps the notion of genius itself needs as much unpacking as does the idea of gender roles.

Thankfully, women these days (at least in the West) have been increasingly freed from many of their culturally imposed shackles and are more able than ever to explore and express their own genius. But this shouldn't obscure the fact that even as women are urged to "lean in," men have scarcely needed comparable encouragement, perhaps because they have long been inspired by their biology to stand out.

The evolutionary basis of homosexuality is one of biology's greatest unsolved mysteries.[11] This is especially true because it is mathematically demonstrable that even small differences in reproductive success can eliminate traits, and individuals who are exclusively homosexual or even bisexual have lower reproductive rates than do exclusive heterosexuals. The long human history of polygamy leads to several interlocking hypotheses. Because polygyny concentrates available women among a small number of men, this leaves an excess of unmated men (discussed earlier as a likely factor contributing to male-imitated violence).

Any time there is a substantial imbalance in the breeding adult sex ratio, we can anticipate that the more abundant and excluded individuals would be—if nothing else—sexually frustrated. The possibility therefore arises that under polygyny, males might look to other males to satisfy their sexual yearnings, and vice versa for females under polyandry. I don't know of any data that speak to this among nonhuman animals—specifically, whether polygynous species experience higher frequencies of male homosexuality than do monogamous ones. Even if confirmed, such a finding would speak to a possible proximate mechanism leading to homosexuality while leaving the ultimate causative mechanism(s) untouched. It would also fail to explain why homosexuality (at least among human beings) is clearly an intrinsic part of the makeup of certain individuals, and not merely a second-best option for otherwise celibate bachelors and bachelorettes.

Among the possible ultimate explanations, several derive directly from polygyny. Homosexual men living in a polygynous society might experience an advantage over their hetero counterparts if they evoked less competition from dominant males, as well as perhaps experiencing less male–male competition generally. After all, they would have little or no motivation to challenge the supremacy and sexual prerogatives of the dominant harem keeper; and, moreover, insofar as they were clearly identified as having a same-sex preference, they wouldn't be perceived as threatening by these dominant harem keepers.

On the other hand, for their frequency to be maintained, genes conducive to homosexuality would have to experience more than a reduced disadvantage; they would have needed to project themselves, somehow, into the future. Among possible mechanisms, the one most clearly associated with polygyny is based on the conjecture that if these men were at least somewhat inclined to bisexuality as well, they would likely have experienced at least some reproductive opportunities, especially because as identified homosexuals, they wouldn't be guarded against to the degree that horny heteros would have been.

There is some evidence that human polygyny generates social instability, an effect that should be greater in proportion as the reproductively excluded men are heterosexual, and, by the same token, diminished insofar as there are comparatively more homosexuals. A report in the Los Angeles Times[12] noted in 2005 that 400 young teenage men had either chosen to leave polygynous Mormon sects in Utah and Arizona or had been driven out.[13] Although it isn't known what proportion of these excluded males were gay, such an emigration not only points to the disruptive effects of polygyny, but also suggests that if a similar process occurred during human evolution, it could have substantially diminished the effective population size of early hominin groups, rendering them vulnerable to conquest by larger groups containing at least some gay warriors.

A related hypothesis maintains that prehistoric human groups containing a substantial proportion of homosexual men might not only have had access to a larger number of warriors but could also have been more successful, man for man, in military conflicts with other groups insofar as the latter experienced comparatively more withingroup conflicts because of the male–male competition that is unavoidable among heterosexual members of a polygynous species. It has been proposed that primitive warfare between competing groups may have been responsible for substantial aspects of human evolution, including, paradoxically, our capacity to cooperate.[14] If early hominin groups containing high numbers of homosexual warriors were unusually cooperative and also less competitive than were overwhelmingly heterosexual groups, then selection—acting at the level of groups—could possibly have overcome the personal selective disadvantage (lower personal reproductive success) experienced by homosexuals. After all, as we have seen, polygyny predisposes men to potentially disruptive and downright violent male–male competition, which could make it comparatively more difficult for fighting groups composed of competitive, bickering, and potentially backstabbing heterosexual warriors to put up a united front against their enemies.

As noted, this mechanism presupposes that human evolution has been driven, at least in part, by selection acting at the level of groups along with the more generally acknowledged potency of individual and gene-level selection. Although so-called "group selection" has enjoyed some increased followership of late,[15] the overwhelming consensus among evolutionary biologists is that it is highly unlikely to be an important cause of evolution among nonhuman animals, for a number of reasons, notably (1) genes are the fundamental unit of selection, because they, not groups, remain intact over generations; and (2) selection within groups will strongly favor traits promoting individual and gene success, which will nearly always swamp between-group selection for altruism.

In the martial scenario suggested previously, heterosexuality equates to "selfish benefit" insofar as success results in more of a breeder's genes being parentally projected into the future, whereas homosexuality equates to "altruism," which benefits the group. Human evolution might represent one of the very few cases in which selection at the level of groups could have been important, especially, perhaps, if highly cooperative warrior groups containing homosexual men were more successful (because more cooperative and cohesive) than their internally squabbling heterosexual opponents.

Although evidence is increasing that the causative biology of lesbianism differs from that of male homosexuality, we can also speculate that polygamy might have provided impetus for lesbianism as well. This is because polygynously mated co-wives would necessarily receive less one-to-one attention from the harem-keeping male, which in turn would advantage those women who could meet some of their social and sexual needs via one or more of the other co-wives; in addition, lesbian women in these circumstances would be less likely to engage in heterosexual infidelity, which, in current polygynous societies, is severely punished, often by death.

Other connections between polyandry and homosexuality are also possible. Thus, reputed male homosexuals would presumably have aroused less suspicion and sexual jealousy, thereby granting bisexual men greater access to any obliging, polyandrously inclined women. At the same time, societies in which polyandry was institutionalized would have experienced an excess of unmated women, which in turn could have favored lesbianism by a similar process—although reversed—as just proposed for polygyny and male homosexuality.

Then there is God-based religion, especially Judaism, Christianity, and Islam. Each of these major monotheistic traditions focuses on a male figure, one who strongly resembles an alpha male at the head of a social group. Sophisticated theologians typically emphasize that their deity lacks a physical body, or rather, somehow transcends physicality. Sometimes—although more rarely—God is conceived as nongendered. Nonetheless, there is little doubt that the great majority of believers imagine a personal god who can be spoken to, who answers prayers, who has strong opinions, and often discernible emotions as well: sad, angry, pleased, displeased, vengeful, jealous, forgiving, loving, and so forth. Not everyone buys into a sky god with a long white beard, a serious and all-knowing mien, capable of rewarding good behavior and punishing bad. But it doesn't take much imagination to recognize that God, as worshipped in most of the world, is remarkably humanoid, widely perceived as a Great, Big, Scary, Willful, yet Nourishing and Protective Guy.

Each of the these traits deserves scrutiny.

Great: Muslims enthusiastically cry "God is great!"; and although the exact words are specific to Islam, the sentiment is universal, as is the apparent need to proclaim it as part of prayer. Actually, at least among monotheists, God is universally seen as not only great but literally The Greatest—in every respect: power, wisdom, goodness, and so forth, just as tribal chiefs and silverback male gorillas would doubtless describe themselves and demand that their subordinates agree.

Big: Dominant alpha harem-keeping males are typically large in stature, and in tribal as well as modern religious traditions, they generally grant themselves ornamentation (ceremonial regalia, especially headgear) that make them seem bigger yet. Moreover, subordinates are expected to emphasize their diminutive status by bowing, kneeling, and in any event keeping their heads lower than that of the king, pope, bishop, dominant gorilla, or chimpanzee. Size—or apparent size—matters, and at least when it comes to issues of dominance, a diminutive god is something of an oxymoron.

Scary: It is dangerous to challenge the status of the alpha harem master. After all, he got there by being not only omnipotent and omniscient, but also omni-destroying—or at least, highly threatening—when crossed. Fear of God is more than a prerequisite for belief in God; the two are nearly identical.

Willful: God generally has very strong opinions and must be obeyed. A truly omnipotent being presumably could orchestrate things as he chooses, but instead—like an alpha male harem master who is currently in charge but who has to constantly guard against intruders or other invaders (in short, against takeovers by other wannabe alphas, i.e., evidenced by the worshipping of other, lesser gods)—he is jealous, vengeful of those who disobey, vigorously prohibiting any backsliding or counterrevolutionary support for competitors.

Nourishing: One way or another, harem keepers are expected to benefit their underlings, often by their success in hunting, warfare, or by successfully orchestrating not only their own fertility (via their wives) but also the flourishing of the other group members. "The Lord said unto Moses, "Behold, I will rain bread from heaven" (New American Standard Bible, Exod. 16:4).

"He provides food for those who fear him; He remembers His covenant Forever" (Psalm 11:5). "It is He who feeds me and gives me drink. "(Koran 26:79). "It is He who made the earth tame for you—so walk among its slopes and eat of His provision" (Koran 67:15).

Protective: "But you, Lord, do not be far from me," implores the psalmist. "You are my strength; come quickly to help me. Deliver me from the sword, my precious life from the power of the dogs. Rescue me from the mouth of the lions; save me from the horns of the wild oxen" (Psalms 22:19–21). And along the way, save me from your own anger, too!

Guy: Just as dominant, alpha harem keepers are male in the world of mammals generally and of primates in particular, gods and their chief representatives on earth are pretty much universally male. It shouldn't be surprising that religious leaders are prone to employing the Great, Big, Scary, Willful, yet Nourishing and Protective Guy notion when seeking to achieve and reinforce their dominance. After all, such an approach serves their own interest insofar as they are able to surf on the psychological bow waves evoked by such imagery because there is much in *Homo sapiens'* evolved psychology that lends itself to these attitudes being readily summoned. To some extent, it takes a willing subject to be led down the path of perceiving God as a dominant, alpha, harem-keeping entity. Our polygamous history has marinated human beings in precisely this tendency.

In *The Future of an Illusion*, Freud speculated that religion came from displacing the assumption of omniscience, omnipotence, and omni-benevolence that, from the infant's perspective, initially characterizes one's parents. As the child grows older, he or she becomes aware of parental inadequacy, whereupon these godlike traits are projected onto an imagined superparent who is perfect, powerful, and just sufficiently removed from day-to-day events to retain his imagined potency and ability to intimidate. And so, once again, God readily emerges as the idealized harem master. There is no particular need in this case to distinguish between God-as-alpha-male and God-as-parent. Both considerations work in pretty much the same way and point in the same direction. In psycho-speak, the outcome is "overdetermined."

Much in the evolutionary psychology of *Homo sapiens* renders our species susceptible to God as portrayed in the Abrahamic religions. We are deeply sensitive to dominance hierarchies and especially to the need to respect the silverback male and his prerogatives. We are subject to sexual impulses that in our evolutionary past contributed to the success of our ancestors but that also risked serious trouble if they were not deployed cautiously; hence, we are endowed with urges that are powerful but that we also intuitively recognize as potentially dangerous to ourselves, especially if they evoke jealous anger from the powerful male.

In his book *The Naked Ape*, zoologist Desmond Morris wrote that religion's "extreme potency" is simply a "measure of the strength of our fundamental biological tendency, inherited directly from our monkey and ape ancestors, to submit ourselves to an all-powerful, dominant member of the group." Flying over Quebec, Canada, in a small plane, one gets the eerie impression that the various towns are inhabited by relatively small creatures, each family occupying a house that is roughly the same size, except for an exceptionally large building—one in each town—that must be the home of a Very Large Person indeed. In addition to its exorbitant size, each "house of God" is notable for its comparatively elaborate architecture; lovely artworks, and the submissive, downright worshipful devotion regularly displayed by the reverential, subordinate troop members who follow—or at least pretend to follow—the rules He has established, and who feel relieved and reassured as a consequence not only of the obeisance in itself but also confident that they have affiliated with a powerful, dominant leader as a source of protection in a fearsome world.

Along these lines, Jay Glass proposed a reworking of the 23rd Psalm, as it might have been conceived by a chimpanzee.[16] Here is my somewhat revised version of his suggestion, more suitable for an early hominin or proto-ape, because modern-day chimps aren't nearly so alpha-male oriented as are current specimens of *Homo religiosum*:

PSALM XXIII	THE HOMININ'S PRAYER
The Lord is my shepherd;	The dominant male is my leader;
I shall not want.	I'll be okay.
He makes me to lie down in green pastures	He helps me get food and water when I need it.
He leads me beside still waters.	He leads me to the best and safest places.
He restores my soul.	He quells my anxiety.
He leads me in the paths of righteousness for his name's sake.	He tells me what to do to avoid getting into trouble, especially with him.
Yea, though I walk through the valley of the shadow of death,	Even though the savannah is full of dangers,
I will fear no evil,	I will fear no competitors,
for Thou art with me	For You are with me.
Thy rod and Thy staff they comfort me.	Your strength and Your vigor they comfort me.
Thou prepares a table before me in the presence of my enemies.	You protect me from other animals (and from yourself).
You anoint my head with oil.	You help me out.
My cup runneth over.	I'm doing pretty well . . . considering.
Surely goodness and mercy shall follow me all the days of my life;	I feel safe in your territory and among friends and relatives as long as I am in your troop;
and I will dwell in the house of the Lord	and I submit and accept your dominance
Forever.	Forever.

Note the powerful sense of awe, combined with the reassuring subordination that believers regularly report in the presence of their God. I suspect that a blackback gorilla male might well report similar feelings in the presence of the dominant silverback. Moreover, those who didn't feel that way, and who expressed their rebellion outright, probably did not leave many descendants. Natural selection has long favored respect for and deference to the dominant harem keeper.

I earlier referred to biologist John Byers's research on pronghorns in which he interprets their extraordinary and seemingly uncalled-for running speed. To reiterate, noting that there aren't currently any superfast predators in North America, Byers hypothesized that the dazzling velocity of pronghorns is due to their being "haunted by the ghosts of predators past." By the same token, human beings are generally haunted by the ghosts of polygyny past, and more specifically perhaps—when it comes to their vulnerability to a particular kind of deity worship—the ghosts of dominant, alpha male, harem holders past.

Unlike the hypothesized connection between polygyny and genius, as well as between polygyny and homosexuality, the suggestion that God has been created in the image of a dominant, harem-keeping alpha male is not unique to this book. It doesn't take a great deal of imagination or biological sophistication to conclude that human beings created God in the image of man (specifically *man*, rather than woman), and not simply because the world's major monotheisms typically refer to Him as a him and not a her. For all the description of God as loving and caring, He also is consistently depicted as homicidally violent, sexually jealous, pushy and demanding, everything that we would expect in a dominant, alpha male, polygynous ape.

In *Alpha God: The Psychology of Religious Violence and Oppression*, Hector Garcia[17] posits that "depictions of the Abrahamic God, and of male gods from religions around the world, reflect the essential concerns of our primate evolutionary past—namely, securing and maintaining power, and using that power to exercise control over material and reproductive resources."

Garcia goes on, noting that

> Male primates struggle for dominance within social groups, using a variety of strategies—fear and aggression among them—to acquire rank status. Rank, in turn, typically confers rewards, which for males includes preferential access to resources such as food, females, and territory. Dominant apes and men have a long history of securing such biological treasure by perpetrating violence and oppression on lower-ranking members of their societies. Once we observe that God, too, is portrayed as having great interest in these kinds of resources, and as securing them through similar means, it becomes increasingly clear that He has emerged as neither more nor less than the highest-ranking male of all.

The traditional monotheistic God meets all the prerequisites, including a whole lot of what biologists call "mate guarding," and no small amount of sexual voraciousness plus acquisitiveness. "God appears to want our women," writes Garcia:

Despite His resplendent powers, He seems to find Himself cloistering women away, policing their sex lives, and punishing the sexual ambitions of lesser males. Such ordinary sexual concerns do not easily reconcile with the notion of an everlasting being, one that requires neither women nor sex to reproduce itself into future generations. Add to this His fixation with sexual loyalty, and we have a dominant male god at His worst behaving like a lustful, jealous, sexually acquisitive male primate.

If, as Garcia and I are suggesting, God is widely conceived through the lens of hominin polygyny, then His basic traits, including His all-too-human sexual appetites, should approach being what anthropologists call a cross-cultural universal. And in fact, they do. As Garcia points out

Zeus was a womanizer, like other powerful males. Krishna, considered a Supreme Being among gods, is a central deity in Hinduism. Krishna was born a cowherder and in early life seduced numerous milkmaids into ecstatic "dances" in the forests. To the dances the milkmaids brought offerings of food, jewelry, their clothing, and, naturally, their bodies. As the story goes, after a stretch of playing hard to get Krishna had sex with all the milkmaids for six long months. The milkmaids fell into deep, ecstatic longing for Krishna, much of which was titillated by his reputation as a powerful warrior. Notably, many of Krishna's sexual conquests were with married women, making their cowherder husbands Krishna's cuckolds in the evolutionary competition for sex. Krishna aptly used his wiles and status as a god toward this end, and the women willingly gave up their lowly cowherder husbands for Krishna, dominant male that he was. But in competition Krishna did not rely on charm alone. When necessary, he slaughtered other dominant males to gain their women and territory. For example, Krishna overthrew and killed his uncle king Kansa, and took Kansa's subjects to establish his own kingdom. Krishna then married the princess Rukmini by abducting her from an arranged marriage with king Shishupala, whom he later killed. Knowing Krishna's status, Rukmini enthusiastically welcomed this new arrangement. Krishna went on to kill Narakasura, a male demon, and in doing so won 16,100 maidens. This was a spectacular prize even by the standards of the most successful despotic kings. Like the gods, there is a great deal of complexity to Krishna's personality. He can show beneficence and kindness if he chooses. But his primate origins predispose him to engage in violence and sexual domination.

The Abrahamic God is less obviously lecherous, but He makes it clear that He is "a jealous God" and one who is especially intolerant if any of His flock dabbles with any other deities. "You shall not make for yourself an idol, or any likeness of what is in heaven above or on the earth beneath or in the water under the earth. You shall not worship them or serve them; for I, the Lord your God, am a jealous God." (Exod. 20:4–5). Unbelievers and those who worship "false idols" are treated with especially violent zeal; like all harem-keepers, he is ruthless in dealing with competition from others who might threaten to displace Him. Moreover, the biblical God personalizes whole cities (notably Babylon and Israel) as women in his life, whose obedience and fidelity are paramount.

God's demands for exclusivity in worship seem to have been taken directly from the playbook of a sexually distrustful dominant male. Insisting that one must "have no other gods before me," is a lot like you must "have no other lovers other than me." Here is God, berating Israel for having "gone astray," in the process conflating the worship of other gods with sexual infidelity:

> "If a man divorces his wife and she leaves him and marries another man, should he return to her again? Would not the land be completely defiled? But you have lived as a prostitute with many lovers—would you now return to me?" declares the Lord. "Look up to the barren heights and see. Is there any place where you have not been ravished? By the roadside you sat waiting for lovers, sat like a nomad in the desert. You have defiled the land with your prostitution and wickedness. . . ." (Jer. 3: 1–2).

Moreover, as with those other dominant polygynous males with which our ancestors had to deal, a cuckolded God is a dangerous one: "And I will judge thee, as women that break wedlock and shed blood are judged; and I will give thee blood in fury and jealousy" (New American Standard Bible, Ezekiel 16:38). The specificity of the threat could have come directly out of the playbook of successful—and violent—polygynists past: "Ye shall not go after other gods, of the gods of the people which are round about you; (For the Lord thy God is a jealous God among you) lest the anger of the Lord thy God be kindled against thee, and destroy thee from off the face of the earth" (Deut 6:14).

Sexual repression of God's followers is also a hallmark of monotheistic religions, just as polygyny is based on restraint on the part of the dominant male's co-wives as well as the other, subordinate and—if they know what's good for them—nonbreeding males. Otherwise, the punishment is to "destroy their altars, break their images, and cut down their groves: For thou shalt worship no other god: for the Lord, whose name is jealous, is a jealous God: Lest thou make a covenant with the inhabitants of the land, and they go a whoring after their gods" (Exodus 34:13).

Garcia argues that sexual repression is

> rooted in mental architecture designed for navigating the social world of our primeval ancestors. In non-human primates, repression is invariably related to rank, occurring when those higher in rank strive to stave off the sexual impulses of those lower. Dominant male humans enact similar strategies, as when kings reign over harems and make eunuchs out of other men, for example. Neither are gods immune from such primate desires and jealousies.

In many different religious contexts we see men and women staving off sexual impulses at the command of the dominant male archetype embodied in God. During Ramadan and Lent millions of Muslims and Catholics (respectively) abstain from sex as a sacrifice to God, as one widely manifested example.

If, as suggested, we are prone as a species to imagine God as a dominant male harem keeper, then it follows that this God—consistent with mammalian male anxieties generally—would be especially concerned about His paternity, with such concern manifesting itself not only in aggressive, threatening behavior toward potential sexual competitors, but also in seeking reassurance that He hasn't been cuckolded.

Accordingly, let's look at one of the best-known, yet insufficiently examined, Biblical passages from Genesis 1:26–28. "Then God said, 'Let us make mankind in our image, in our likeness, so that they may rule over the fish in the sea and the birds in the sky, over the livestock and all the wild animals, and over all the creatures that move along the ground.'" Here we have the purported—but potentially contested—offspring being proclaimed by the Chief Polygynist to be legitimate Chips off the Old Divine Block. And God further announces that because of this confirmed connection, His children are the legitimate inheritors of, well, everything.

The creation myths of many non-Western religions include description of how people were generated, commonly out of some divine body part. Judeo-Christian tradition is perhaps the most explicit and complete in this regard, known in Catholic doctrine as the principle of *Imago Dei*, whereby God essentially reproduced Himself in humanity (although presumably leaving out a few components). Here is Genesis once again: "When God created mankind, He made them in the likeness of God" (5:1).

Among the most influential Catholic theologians are Saints Augustine and Thomas Aquinas; and once again I thank Hector Garcia, this time for pointing out the following example of Aquinas quoting Augustine on the implications of *Imago Dei*: "[T]he body of man alone among terrestrial animals is not inclined prone to the ground, but is adapted to look upward to heaven, for this reason we may rightly say that it is made to God's image and likeness, rather than the bodies of other animals." By a similar zoo-logic, we might conclude that those flatfish commonly known as flounders (members of the suborder *Pleuronectoidei*) are even more reflective of *Imago Dei* because although they hatch with eyes on either side of their heads like most vertebrates, one eye migrates to join the other during embryologic development so that unlike other creatures (including *Homo sapiens*), both of their eyes end up in the dorsal region of their bodies, making them not only capable of looking upward to heaven, but unable to avoid doing so.

In any event, many Bible passages assert not only a family resemblance but God's direct paternity. For example, "Ye are the children of the Lord your God" (Deut. 14:1); "All of you are children of the most High" (Proverbs 82:6); "Ye are the sons of the living God" (Hosea 1:10); "We are the offspring of God" (Acts 17:29). Furthermore, this relationship is strongly colored with recognizable human emotion, notably reassurance that God feels paternal love, specifically in recognition of His followers' filiation:

"See what great love the Father has lavished on us, that we should be called children of God!" (1 John 3:1).

Imago Dei underpins the widespread insistence—found in nearly every religion—that followers are "God's children." More accurately, however, *Imago Dei* is a case of *Imago Homo* in which God was created in the image of man—or perhaps, *Imago Polygynosus*. Whatever the appropriate Latin phrase, the biological reality is that religious traditions tend to emphasize precisely the continuity (genetic transubstantiated into theologic) that evolutionary considerations would predict.

Evolutionary geneticists are quite aware, for example, that parent and child have a genetic relationship of 50%, with this connection diluted by 50% in each generation. In the Christian tradition, Jesus derives his authority from being literally the Son of God (and by a nifty twist of theology, the Son of Man as well). If God is the father, and Jesus the son, then the only quasi-genetic route for the next generation—that is, regular *Homo sapiens*—to the former is via the latter. Jesus is clear on this (albeit minus the gene-based arithmetic) when, in John 8:42–47, he remonstrates with the Jews over their refusal to accept that God is his literal father:

> "We are not illegitimate children," they [the resisting Jews] protested. The only Father we have is God himself." "If God were your Father [replies Jesus] you would love me, for I have come here from God. I have not come on my own; God sent me. Why is my language not clear to you? Because you are unable to hear what I say. You belong to your father, the devil, and you want to carry out your father's desires. . . . Whoever belongs to God hears what God says. The reason you do not hear is that you do not belong to God."

In short, if you are legitimate offspring of God, you must accept what I say, because—being just one generation removed and therefore closer than you to God—I am the necessary (biological) intermediary: "No one comes to the Father, but by me" (John 14:6). And if you aren't the descendants of God then you "belong" to Satan, which is to say, to some other male. Bad choice!

As we have already seen, "belonging" to another male is precarious, especially in a polygynous species that is liable to infanticide. And once again, Judeo-Christian tradition conforms so closely to sociobiological expectation that we might almost wonder if someone had "fudged" the data. The Old Testament in particular brims with horrific exhortations of infanticide, consistent with biological reality if not current morality, and predictably evoked when the ancient Israelites conquer an unrelated tribe that worships (i.e., is genetically or otherwise connected to) other gods: for example, "Now therefore kill every male among the little ones, and kill every woman who has known man lying with him. But all the young girls who have not known man lying with him, keep alive for yourselves" (Num. 31:17–18).

A common characteristic of religious traditions—all of them, not just the Abrahamic Big Three—is the relegation of certain activities to human beings but with direct divine authority. When Jesus proclaims "render unto Caesar that which is Caesar's," he is facilitating the smoothly functioning secular success of his representatives on

Earth, those who have been granted the Chief Polygynist's proxy and who often take lascivious advantage of it. On the other hand, when they take vows of chastity and adhere to them, they are especially pleasing to God, having joined His harem. Thus, nuns become "brides of Christ," whose fidelity is thereby emphasized. At the same time, the Catholic church demands (although does not always get) male chastity as well, at least on the part of priests, who basically are supposed to give up their sexuality in the service of a more dominant male. "Clerics are obliged to observe perfect and perpetual continence," as stated in Canon 277, "for the sake of the kingdom of heaven and therefore are bound to celibacy which is a special gift of God by which sacred ministers can adhere more easily to Christ with an undivided heart and are able to dedicate themselves more freely to the service of God and humanity."[18]

According to Paul, by forgoing sexual relations with women, priests forgo pleasing their wives so as to please God: "The unmarried man is anxious about the affairs of the Lord, how to please the Lord; but the married man is anxious about worldly affairs, how to please his wife, and his interests are divided. (1 Cor. 7:32–34). Not surprisingly, it is especially pleasing to any dominant polygynist when his women are chaste and other men are celibate. As Paul also puts it, "better to marry than to burn" . . . with sexual desire (I Cor. 7:9), but better yet to refrain from challenging the Big Guy's prerogatives, which includes making sure you do nothing to threaten His confidence of paternity.

Harem-master males (regardless of species) are often at great pains to constrain the sexual ambitions of their subordinates; and so, there are few religions in which God is portrayed as favoring sexual license, and many in which acolytes are expected to practice abstinence, with virginity and celibacy being especially prized.

Not surprisingly, numerous religious traditions—the Abrahamic Big Three most especially—maintain that God strongly disapproves of various sexual practices, not just adultery (although that, too, big time). According to Garcia, "He has been described in the Abrahamic faiths as having personal distaste for extramarital sex, homosexuality, prostitution, oral sex, anal sex, masturbation, revealing clothing, and even sexual thoughts." Sexual restraint is a terrific way to avert jealous anger on the part of any dominant harem-keeper.

Genital mutilation also abounds in some traditions—notably certain ethnic groups within Islam and all of Judaism—in the former case as a way of controlling women's sexuality and in the latter, as testimony to the commitment of His followers. It is interesting to note that the word "testimony" derives from the Latin translation of testicle. In ancient Rome, two men swearing allegiance would hold each other's testicles, presumably to indicate mutual confidence (which may have been confirmed insofar as neither squeezed too hard!).

A deep question arises, one that should be troubling to believers but generally is not. Given that God is ostensibly infinite in power and extent, why should it matter to such a deity whether He is granted deference, obedience, a worshipful monopoly when it comes to subordination; kept free from criticism or doubt; reserved the choicest virgins and receiving the most valuable sacrifices? This doesn't make sense except insofar as these gods are in fact distorted reflections of our own evolved psychology, which has elaborated primitive hominin tendencies for obeisance to dominant yet anxiety-ridden, harem-holding males who are powerful and potentially deadly, jealous and vengeful, especially when their prerogatives are challenged.

It is at least worth considering that the evolved human brain—for all its wonders— also contains a mammalian component that has been largely (and naturally) selected in an environment of male-dominated polygyny, along with more subtle, female oriented polyandry. As a consequence, we are predisposed not only to overt polygyny plus covert infidelity and polyandry, but also to a familiarity with and inclination to participate in systems of social deference and followership associated with a dominant, alpha male god, with that god created in the image of a dominant, alpha male polygynist. Not a pretty picture, but as Darwin noted in *Sexual Selection and the Descent of Man*, "We are not here concerned with hopes or fears, only with truth as far as our reason permits us to discover it."

The preceding hypotheses regarding genius, homosexuality, and monotheistic religions suggest that these phenomena developed as possible by-products of human polygyny and polyandry. They are, of course, speculative, and may or may not turn out to be productive scientific leads; indeed, they will probably prove easier to pose than to evaluate. In the spirit of honest inquiry, however, and at the risk of hypothesis overkill, it seems worthwhile to suggest some additional possible correlates of our penchant for polygamy.

We've seen that polygyny in particular is a direct outgrowth of high male variance in reproductive success. Is it possible that income and socioeconomic inequality— not just the tolerance of such differences, but the establishment of circumstances underlying their very existence—is associated with a polygynist mindset, whereby some individuals accrue "more" than others? Certainly it is more than their "fair share." It seems at least possible that capitalism's fondness for "more," the acquisitiveness that underlies not only much human social and economic activity but currently threatens the worldwide ecosystem, is a consequence of an evolutionary process that has—at least in the past—rewarded those who obtain "more" (a bottom line of reproductive success projected onto social and material success).

Finally, there is yet another potential consequence of polygamy, one that is so obvious that (like male-oriented violence), it is easy to overlook: even now, in the 21st century, many social institutions continue to reflect a harem-haunted world. The overwhelming majority of corporations, government institutions, institutions of higher learning, and indeed, social organizations in general are hierarchically organized and male dominated, often with a pyramidal structure in which a small number of men are at the top, with high-ranking immediate subordinates often male as well—albeit with increasing numbers of female subalterns as the hierarchy is descended—and with numerous subordinate males and the equivalent of co-wives at the lowest levels.

Insofar as these and other traits—most of them regrettable—are attributable to the ghosts of polygynists past, there is all the more reason to inquire into how our evolutionary past interfaces with our cultural present. We'll do this in the remaining chapters.

Notes

1. Oscar Wilde, quoted in Montagu, A. (1953). *The natural superiority of women.* New York: Macmillan.
2. This, incidentally, is why evolutionary psychology (or its older identifying moniker, sociobiology) is, rather than inherently racist as some critics contend, actually one of science's most potent antidotes to racism.
3. Konner, M. (2015). *Women after all.* New York: W. W. Norton.
4. Miller, G. (2001). *The mating mind.* New York: Anchor.
5. Brooks, R. (2011). *Sex, genes & rock 'n' roll: How evolution has shaped the modern world.* Sydney, Australia: University of New South Wales Press.
6. Hawkes, K. (1990). Why do men hunt? Benefits for risky choices. In E. Cashdan (Ed.), *Risk and uncertainty in tribal and peasant economies.* Boulder, CO: Westview Press.
7. Stanford, C. B. (2001). *The hunting apes: Meat eating and the origin of human behavior.* Princeton, NJ: Princeton University Press.
8. de Waal, F. (1989). *Peacemaking among primates.* Cambridge, MA: Harvard University Press.
9. de Waal, F. (1984). Sex differences in the formation of coalitions among chimpanzees. *Ethology and Sociobiology, 5,* 239–255.
10. Allen, A. (2014). Chess grandmastery: Nature, gender, and the genius of Judit Polgár. JSTOR Daily. http://daily.jstor.org/chess-grandmastery-nature-gender-genius-judit-polgar/?utm_source=internalhouse&utm_medium=email&utm_campaign=jstorda ily_10292014&%E2%80%A6.
11. Barash, D. P. (2013). *Homo mysterious: Evolutionary puzzles of human nature.* New York: Oxford University Press.
12. Kelly, D. (2005, June 13). Lost to the only life they knew. *LA Times,* p. A1.
13. I am grateful to Sebastian Donovan, who informed me of this.
14. Alexander, R. D. (1971). The search for an evolutionary philosophy of man. *Proceedings of the Royal Society of Victoria, 84,* 99–120.
15. Wilson, D. S., & Wilson, E. O. (2007). Rethinking the theoretical foundation of sociobiology. *Quarterly Review of Biology, 82,* 327–348.
16. Glass, J. D. (2007). *The power of faith: Mother nature's gift.* Corona del Mar, CA: Donington Press.
17. Garcia, H. A. (2015) *Alpha God: The psychology of religious violence and oppression.* Amherst, NY: Prometheus.
18. Once again, I thank Hector Garcia's marvelous book for making me aware of this material.

8

The Hare and the Tortoise, Redux

The ape-man exultantly threw his club (actually, the leg bone of an early mammal) into the air, and as it spun, it morphed into an orbiting space station. In this stunning image from the movie *2001: A Space Odyssey*, millions of cinemagoers saw the human dilemma in microcosm. We are unmistakably animals, yet we also behave in ways that transcend the merely organic. Ape-men and ape-women all, we are the products of biological evolution—a Darwinian process that is both slow and altogether organic—yet at the same time we are enmeshed in our own cultural evolution, which, by contrast, is blindingly fast and proceeds under its own rules.

Like it or not, we have inherited a range of biologically mandated predispositions, inclinations, and tendencies with which we are stuck. Some, such as the anatomy and physiology that keeps our hearts beating, our lungs breathing, our livers detoxifying, and so forth are life generating and accordingly, not subject to critique (except when they fail). Others, such as our underlying polygamy, are troublesome, sometimes extremely so, especially when they interact with culturally generated rules and expectations. The good news is that unlike breathing or circulating our blood, many of these behavioral inclinations are just that: inclinations. Although they cannot be altogether erased, there is good reason to be confident that they can be transcended—once we acknowledge their existence and can therefore arm ourselves against their tyranny.

But first, a bit about the curious situation in which *Homo sapiens* finds itself.

While the cinematic ape-man's club traveled through air and ultimately into outer space, director Stanley Kubrick collapsed millions of years of biological and cultural evolution into five seconds. My point, however, is that this wasn't simply a cinematic trick. We are all time travelers, with one foot thrust into the cultural present and the other stuck in the biological past. One aspect of that past, the subject of this book, is our confused and confusing biological history when it comes to how we connect with other human beings over something no less important evolutionarily than breathing air or circulating blood: reproducing.

As purely biological creatures, we are neither more nor less "special" than our fellow organic beings. As cultural creatures, however, we are extraordinary indeed. We compose symphonies, travel to the moon, and explore the world of subatomic particles. But at the same time we are unique among living things in being genuinely uncomfortable in our situation. This should not be surprising because even though our cultural greatness must have somehow sprung from our organic beast-liness, the two processes (organic and cultural) have become largely disconnected, and as a result, so have we: from ourselves, each other, and often, our own genuine self-interest.

The little hyphen in ape-man and ape-woman is the longest line conceivable, con-necting two radically different worlds, one biological and the other cultural. Imagine two people chained together; one a world-class sprinter, the other barely able to hob-ble. Now, imagine that they are both expected to run as quickly as possible. The likely outcome includes a bit of tension all around.

To understand why biological and cultural evolution can experience such con-flict (despite the fact that they both emanate from the same creature), consider the extraordinarily different rates at which they proceed. Biological evolution is unavoid-ably slow. Individuals, after all, cannot evolve. Only populations or lineages do so. And they are shackled to the realities of genetics and reproduction because organic evolution is simply a process whereby gene frequencies change over time, an ongoing but ponderous Darwinian event in which new genes and gene combinations are eval-uated against existing options, with the most favorable making a statistically greater contribution in succeeding generations. Accordingly, many generations are required for even the smallest evolutionary step.

By contrast, cultural evolution is essentially Lamarckian, and astoundingly rapid. Acquired characteristics can be "inherited" in hours or days, before being passed along to other individuals, then modified yet again and passed along yet more—or dropped altogether—everything proceeding in much less time than a single genera-tion. Take the computer revolution. In just a decade or so (less than an instant in biological evolutionary time), personal computers were developed and proliferated—also modified, many times over—such that they are now part of the repertoire of most technologically literate people. If, instead, computers had "evolved" by biologi-cal means, as a favorable mutation to be possibly selected in one or even a handful of individuals, there would currently be only a dozen or so computer users instead of several billion.

Just a superficial glance at human history shows that compared with the rate of biological change, the pace of cultural change has been not only rapid, but if any-thing, the rate of increase in that change seems itself to be increasing, generating a kind of logarithmic curve. Today's world is vastly different from that of a century ago, which is almost unimaginably different from 50,000 years ago—not because the world itself or the biological nature of human beings has changed, but because such cultural creations as fire, the wheel, metals, writing, printing, electricity, inter-nal combustion engines, television, and nuclear energy have arisen and then been "inherited" at blinding speed.

Try the following *Gedanken* experiment. Imagine that you could exchange a new-born baby from the mid-Pleistocene era—say, 50,000 years ago—with a 21st-century newborn. Both children—the one fast forwarded no less than the other brought back in time—would doubtless grow up to be normal members of their society, essentially indistinguishable from their peers who had been naturally born into it.

A Cro-Magnon infant, having grown up in 21st-century America, could very well end up reading this book, perhaps on an e-reader or computer screen of one sort or another; whereas similarly, the offspring of today's technologically comfortable adults would fit perfectly if he or she grew up in a world of mastodon skins and chipped stone axes. But switch a modern human adult and an adult from the late Ice Age, and there would be Big Trouble, either way. Human biology has scarcely changed in tens of thousands of years, whereas our culture has changed radically.

Admittedly, our capacity for culture is itself a product of our biological evolution, and yet, this is no guarantee that the two must proceed in synchrony. If anything, the opposite is more likely because culture, like an errant and headstrong child—or Frankenstein's monster—has become disconnected from its biological moorings and has pretty much developed a momentum of its own, proceeding nearly independently of the biological process that originally spawned it. This is because cultural evolution has the capacity to take off on its own, to mutate, reproduce, and spread with such speed as to leave its biological parent far behind in the dust. In theory, the two might still be pointed toward the same ends, but biological evolution remains shackled by genetics—and thus, it lumbers along at the pace of a tortoise, never faster than one generation at a time, and nearly always much slower than that—while cultural evolution plays by its own rules, which often means a mad dash, like a hare.[1] There isn't even much reason to expect the two to be headed in the same direction.

To change the metaphor, two huge continents have drifted apart and now these great tectonic plates, culture and biology, grind together. The results range from nearly trivial squeaks and wriggles, such as our troublesome sweet tooth or some of our personal peccadilloes, to the most portentous quakes, including nuclear war, environmental abuse, and overpopulation—while in-between lie a host of middle-sized tremors such as personal alienation, family dysfunction, and other conflicts between our biologically generated inclinations and our culturally mandated expectations and restraints. The conflict between culture and biology, the peculiar race between hare and tortoise, is an event of paradoxical proportions, ranging from the seismic to the microscopic, from whole societies (indeed, the whole planet and its past, present, and future) to individual people and their likes and dislikes.

Before considering the multifaceted connection between our biological and cultural evolution in the realm of polygamy, monogamy, and in-between, let's briefly consider violence and aggression, because this, after all, was what our cinematic ape-man was doing when so adroitly portrayed on film. Moreover, it is an issue whose outcome is even more fraught and consequential than our sociosexual inclinations. The history of "civilization" is, in large part, one of ever-greater efficiency in killing: with increasing ease, at longer distances, and in larger numbers. Just consider the "progression" from club, knife, and spear, to bow and arrow, musket, rifle, cannon, machine gun, battleship, bomber, and nuclear-tipped ICBM. At the same time, the human being who creates and manipulates these devices has not fundamentally changed at all. Considered as a biological creature, *Homo sapiens* is poorly adapted to kill. The reality is that with our puny nails, nonprognathic jaws and laughably tiny teeth, a human being armed only with his or her biology is hard-pressed to kill just one fellow human, not to mention hundreds or millions. But culture has made this not only possible but easy.

Animals whose biological equipment make them capable of killing each other are generally disinclined to do so. Eagles, wolves, lions, and crocodiles have been outfitted by organic evolution with lethal weaponry and, not coincidentally, they have also been provided with inhibitions when it comes to using them against fellow species members. (This generalization was exaggerated in the past. Today, we know that lethal disputes, infanticide, and so forth do occur, but the basic pattern still holds: rattlesnakes, for example, are not immune to each other's venom; yet when they fight, they strive to push each other over backward, not to kill.)

Because we were not equipped, by biological evolution, with lethal weaponry, there was little pressure to balance our nonexistent organic armamentarium with behavioral inhibitions concerning its use. One reason why guns are so dangerous is that the lethal consequence of a very small movement—curling a finger around a trigger, with barely a few ounces of pressure—are magnified, by superb technology, into violent acts of dreadful consequence. If, by contrast, we had to live—and die—by the application of direct, biological force alone, there would be far more living and less dying.

The disconnect between culture and biology is especially acute in the realm of nuclear weapons. At the one-year anniversary of the bombing of Hiroshima, Albert Einstein famously noted that "the splitting of the atom has changed everything but our way of thinking; hence we drift toward unparalleled catastrophe."

He might have been talking about musk oxen. These great beasts, like shaggy bison, occupying the Arctic tundra, have long employed a very effective strategy when confronted by wolves, who were pretty much their only enemies. Musk oxen respond to a wolf attack by herding the juveniles into the center, while the adults face outward, arrayed like the spokes of a wheel. Even the hungriest wolf finds it intimidating to confront a wall of sharp horns and bony foreheads, backed by a thousand pounds of angry pot roast. For countless generations, their antipredator response served the musk ox well. But now, the primary danger to musk oxen is not wolves but human hunters, riding snowmobiles and carrying high-powered hunting rifles. Under this circumstance, musk oxen would be best served if they spread out and hightailed it toward the horizon, but instead they respond as previous generations have always done, forming their trusted defensive circle and easily slaughtered.

The invention of the snowmobile and the rifle have changed everything but the musk ox way of thinking; hence, they drift toward unparalleled catastrophe. They cling to their biology, even though culture—our culture—has changed the payoffs. Human beings also cling to (or remain unconsciously influenced by) their biology, even as our culture has dramatically revised most payoffs for ourselves as well. That musk ox-like stubbornness is evident when it comes to thinking—or not thinking—about something less consequential for the planet than nuclear weapons, but more immediately relevant to the lives of most people: our own mating biology.

This doesn't mean, however, that our situation is hopeless. Human beings are unique in the biological world in our ability to say "No" to many of our evolved promptings. We can't voluntarily stop our hearts and can only hold our breath for a limited time before a biological override kicks in. It may be difficult, but we can embrace nonviolence, denuclearization, even—one hopes—a worldwide ethic of what the late Carl Sagan called "basic planetary hygiene."

W hen it comes to polygamy, as with many other conflicted and confusing aspects of our behavioral inheritance, we are both biological tortoise and cultural hare, rolled into one body. Some of our most difficult challenges derive from the mismatch between our biologically generated proclivities (including, but not limited to polygamy) and our culturally produced opportunities and situations, confronted by the yearning of women to be autonomous actors—subjects rather than objects, as Simone de Beauvoir cogently put it[2]—yet forced to deal with certain biological propensities and confounded by cultural rules and roles (the "double standard" not least). By the same token, men need to deal with what is often a straitjacket imposed on them by a combination of social expectations and evolutionary hangovers associated with their atavistic role as aspiring harem masters, as well as being subordinate males needing to defer to "the Man."

Sometimes the hare and the tortoise are remarkably similar, or maybe unremarkably. This is because if culture doesn't track biology, at least to some extent, the result is either an impossible situation—imagine a cultural tradition that required people to stop eating altogether—or one unlikely to persist, such as the Shakers' denial of sexual activity among its members. In any event, a mainstay of evolutionary research into human behavior involves making predictions about behavior we expect to occur given what we know about natural selection, or—somewhat less impressive—observing certain behaviors and then attempting to assess whether these behaviors are consistent with expectations based on our understanding of evolution.

In some cases, cultural practices (the hare) have developed so quickly that they present our biology (the tortoise) with situations for which the latter is unprepared. The technology of violence in general, and of nuclear weapons in particular, are notable examples. In other situations, cultural practices act on our biology to generate a situation in which the predispositions of our inner tortoise are exaggerated. A good example is the "claustration" of women, defined as "their concealment, beginning at or before puberty in areas easily controlled by their natal or affinal kin, to prevent contact with males who are potential sexual partners, whether strangers or relatives."[3] The preeminent student of claustration is anthropologist Mildred Dickemann, from whom this description is taken. Dickemann goes on to point out that

> The *se*clusion of women is in fact the *ex*clusion of men. . . . The human socialization process makes claustration far more a matter of values than of force, values shared by the overwhelming majority of both sexes in the contexts in which it occurs. Intersexual dining and socializing between adults are tabued, with exceptions in some cases for certain categories of kin, and public religion, especially in its most formal, prestigious aspects, is restricted to men. The segregation encourages the construction of architectural devices from the

simples of curtains or screens to the most elaborate of walled courtyards and recessed women's apartments; it may involve special modes of transportation in enclosed vehicles or palankeens, and guarding of the women's quarters by night-watchmen, eunuchs, elderly male, or female servants. Slaves, servants, or male members of the kin group may take over public chores such as marketing or water carrying, elsewhere assumed by women.

The best known manifestation of claustration to Westerners is the widespread Islamic practice of veiling, which, as Dickemann notes, "covers, in increasing intensity, the nape of the neck, the hair and top of the head, the forehead, the chin, the face, and finally the whole body from top to toe, the greatest degree of covering being assumed in public and in the presence of certain categories of males."

It isn't at all clear—moreover, it seems highly unlikely—that Islamic men are more sexually avaricious than males from other religious traditions, nor that Islamic women are especially alluring. (This would seem to be a question that could readily be answered by empirical, cross-cultural research.) It is presently obscure why such extreme measures developed in some societies rather than in others. At the same time, it is likely that once the tradition of veiling parts of women's bodies became widespread—whatever the reason—this set in motion a process whereby a glimpse of forbidden body parts became particularly arousing, so that a positive feedback could have been initiated, with enhanced arousability of men leading to increased restrictions on women,[4] primarily orchestrated and enforced by men—who, as we have seen, have much to lose if "their" women are impregnated by someone other than themselves.

Support for veiling in Islam comes largely from a single Surah (34:59) of the Koran: "O prophet! Tell thy wives and thy daughters and the women of the believers to draw their cloaks close round them when they go abroad. That will be better, so they may be recognized and not annoyed." It can be claimed that such a prescription is intended to protect the women in question, for their own benefit, "so they may be . . . not annoyed." Indeed, I have heard from several women, exchange students at the University of Washington from Muslim countries, that they feel less safe and less valued on the streets of Seattle, if they partake of the Western habit of comparatively revealing dress codes, than back home, where their bodies are obscured.

But don't forget the earlier Koranic mention that veiling is designed "so they may be recognized." In this usage, "recognized" definitely does not refer to being identified personally, since if anything, veiling interferes with such recognition. The idea instead is that the women in question will be recognized—that is, publicly acknowledged—*as the property of, or under the protection of* (take your pick) a man. There is undoubtedly no small amount of patriarchal control and male–male competition herein intended and expressed. Further evidence for this is the fact that historically in Islam, veiling typically wasn't required and often was explicitly prohibited for women who were professional entertainers, prostitutes, or slaves. Concern with the degree of a woman's claustration has varied with their socioeconomic status (more accurately, that of their male relatives).

According to Elizabeth Cooper, a British student of Islamic and Oriental women writing a century ago,[5] in Northern India

> There is a saying that you can tell the degree of a family's aristocracy by the height of the windows in the home. The higher the rank, the smaller and higher are the windows and the more secluded the women. An ordinary lady may walk in the garden and hear the birds sing and see the flowers. The higher grade lady may only look at them from her windows, and if she is a very great lady indeed, this even is forbidden her, as the windows are high up near the ceiling, merely slits in the wall for the lighting and ventilation of the room.

Foot binding in traditional China is a parallel example. In this practice, no longer permitted, the feet of young girls were severely wrapped, thereby causing them to be grotesquely deformed so that the victims couldn't walk unaided. Although traditional Chinese accounts describe how husbands adored the "tiny, perfect, moon-shaped feet" of their wives, it seems likely that such adoration came as a consequence of a cultural practice that served biological ends: restricting the independence, and thus the sexual options, of the women involved.

Dickemann points out that it is precisely in those societies that were the most rigidly stratified

> with the greatest extremes between the rich and poor, in which court and palace enjoy unimagined luxury while the masses live on a level always close to starvation, in which large numbers of beggars, outcasts, floater males, and celibates exist at the bottom while intense polygyny in the form of secondary wives, concubines, and harems occurs at the top that the most extreme forms of claustration, veiling and incapacitation of women occur.

Given what we have already learned about the biologically imposed reproductive limitations on women as compared to men, it may seem peculiar that super-high-status families should be so exceptionally concerned about the chastity of their daughters. Bear in mind that super-high-status men had numerous sexual/reproductive avenues available to them, in the form of prostitutes, temporary liaisons, concubines, and so forth. At the same time, there was nonetheless considerable competition to be the preferred or primary wife of a high-status man, and to maximize the amount of paternal resources such a woman (and her offspring) would receive. As a result, the interests of a high-status husband and those of his wife's—or future wife's—family, tended to coincide. So, too, the reproductive interest of the cloistered woman herself, although a strong case can be made that the happiness and subjective well-being of such a "cared for" woman may have been another matter entirely.

Claustration and foot binding are examples of culture enhancing biology in ways that seem ultimately detrimental for the people involved (certainly, the women, although perhaps not their families, or even their genes). In other cases, specifically monogamy versus polygamy, things appear to have been reversed, with culture acting as a brake on the troublesome excesses of our biology. There is little doubt that monogamy is "unnatural," or at least, that people are predisposed to be polygamous (polygynous or polyandrous, depending on their gender), but are often constrained by cultural traditions that prescribe monogamy. Such inhibitions on our "natural" tendencies may be troublesome, but I shall argue that "natural" is a far cry from "good"; and therefore, human flourishing is sometimes well served by cultural requirements and prohibitions.

Probably the best account of this general situation comes from Freud, who, in *Civilization and Its Discontents*, pointed out that people are equipped with an *id*—a ravening core of self-centeredness and uninhibited, often aggressive sexuality—that must be inhibited for civilization to be maintained. The result, he argued, is an array of neuroses, those "discontents" that are demanded by civilization. Troublesome as they often are, it is a cost that human beings are well-advised to pay, given the far greater costs of unrestrained ids leading to a loss of civilized behavior.

Martin Luther King Jr. famously preached that "The arc of the moral universe is long, but it bends towards justice."[6] The arc of biological and cultural evolution is also long, and although it isn't at all clear that it bends toward justice, it does appear—maybe this is a letdown—that it bends toward monogamy. Although museum displays showing Australopithecines typically show one male with one female, fossils show substantial sexual dimorphism, considerably more than is true of modern *Homo sapiens*. There is some debate as to the relative sizes of male and female Australopithecines, with current consensus being that the male to female body weight ratio was 1.52 (among modern humans it is 1.2).[7] So the likelihood is that our ancestors were if anything more polygynous than we are today. Cultural evolution has undoubtedly moved toward monogamy, and maybe our biological evolution will eventually catch up.

This isn't simply a pious hope. Polygyny as an ongoing social condition, albeit buttressed by biology, isn't only the result of certain factors (notably male–female differences when it comes to variance in reproductive success as well as confidence of relatedness to offspring): it also establishes a situation in which the consequences of polygyny are not only maintained but enhanced, for example, male violence, sexual jealousy combined with interest in multiple sex partners. On the other hand, the longer monogamy persists—and the evidence is that despite its current prevalence it is a relatively recent sociocultural innovation—the more selection would favor biological traits conducive to it, insofar as cultural institutions work to disadvantage men who act out their polygynous inclinations as well as women who do the same with polyandry.

We have already seen that men are the primary losers under polygyny, although women, too, are typically worse off under its influence; the only consistent beneficiaries are those relatively few men who end up as harem masters. In addition, the interests of men and women are especially likely to diverge under polygyny (as well as under promiscuity) compared to monogamy. In one notable laboratory study involving fruit flies, these animals—which normally tend toward promiscuity—were kept for many generations in monogamous conditions. In their natural, nonmonogamous state, male flies transmit to females a chemical cocktail that restricts their survivorship as well as their mating opportunities. But after experiencing a multigenerational history of monogamy, "male behavior and seminal fluid proteins evolved to enhance rather than reduce female survival relative to promiscuous controls, and the fitness of monogamous males declined in parallel when they were placed back in their promiscuous ancestral population." In addition "females from the monogamy treatment evolved to be less resistant to male-induced harm when they were placed back in their promiscuous ancestral populations."[8] The research report describing these findings was titled "Dangerous Liaisons," referring to the situation—for females—in the absence of monogamy.

Although fruit flies are far removed from human beings, there may be a valuable lesson here, and not just for entomologists and evolutionary geneticists. Polygamy—both polygyny and polyandry—really does lead to a kind of "war between the sexes," whereas monogamy leads to shared interests. Although polygamy is to some degree "in our genes," that doesn't necessarily make it good, if by "good" we focus on its ethical consequences and if we assume that the arc of such consequences bends—or should bend—toward male–female equality and mutual flourishing.

Earlier in this chapter, when discussing the biological tortoise and the cultural hare with regard to violence and war making, culture emerged as a kind of villain, hijacking the naïve and less sophisticated tortoise, who was unprepared for the latter's deadly, high-tech machinations. Such a perspective, although accurate when it comes to the apparatus of killing, is itself naïve in a broader context, if only because human beings have gone so far down the road of cultural innovations that we literally cannot turn back, nor would it be appetizing to do so. No sane person would turn away from our brave new world that provides antibiotics, protection from predators, food and shelter for millions, opportunities for self-expression, and personal well-being unimaginable to our Pleistocene ancestors.

Jean-Jacques Rousseau, so terribly wrong in his description of the solitary, healthy, unimpeachably happy "noble savage," has also been immensely influential when it comes to his embrace of "natural" as tantamount to "good." In many ways, Rousseau's vision has been useful, a corrective alternative to the modern world in which natural environments are increasingly imperiled; foods are commoditized and adulterated with pesticides, herbicides, and growth hormones; relationships are subordinated to "social media"; and so forth. But in his idealization of wide-open, uncommitted, deliriously happy sexual relations—along with his demonization of society as responsible for most human ills—Rousseau wasn't only mistaken but responsible for much subsequent misunderstanding.

After writing his breakthrough essay known as the *Second Discourse on Human Inequality*, Rousseau sent a copy to Voltaire, hoping for the older man's imprimatur. Voltaire wrote back

> I have received, Monsieur, your new book against the human race. . . . You paint in very true colors the horrors of human society; . . . no one has ever employed so much intellect to persuade men to be beasts. In reading your

work one is seized with a desire to walk on four paws. However, as it is more than sixty years since I lost that habit, I feel, unfortunately, that it is impossible for me to resume it.[9]

Voltaire's sarcasm has not seriously impeded widespread acceptance of Rousseau's interpretation of the human condition. In some ways, Rousseau anticipated the hare-and-tortoise perspective described previously, although without explicitly identifying the twin processes of cultural and biological evolution. Yet when it comes to matters of human sociosexuality—in which the upstart cultural hare causes trouble for the perseverating biological tortoise—the situation is largely reversed. Polygamy is our tortoise, monogamy our hare. And there is much to be said for the latter as a way of ameliorating the excesses of the former.

Sometimes, Mother Nature herself just isn't very nice.

For many, myself included, criticizing nature doesn't come (how else to say it?) naturally. My own preferred recreational activities—hiking, climbing, snorkeling, birding, and riding horses—embed me in nature. I have surrounded myself with animals of all sorts, and I try to avoid consuming pesticides, herbicides, and the antibiotics and hormones to which industrial agriculture has become addicted. My wife and I were delighted when a natural foods supermarket recently opened within a mile of our home, and we patronize it almost exclusively.

Nonetheless, in resisting many things that are "unnatural"—such as nuclear weapons, global warming, chemical pollution, habitat destruction—while also honoring, respecting, defending, validating, supporting, admiring, and nearly worshipping many things that are natural (sometimes even *because* they are natural), it is all too easy to get carried away, to forget that much in the world of nature is unpleasant, indeed odious. Consider typhoid, cholera, polio, Ebola, or AIDS: what can be more natural than viruses or bacteria, composed as they are of proteins, nucleic acids, carbohydrates, and the like? Do you object to vaccination? You'd probably object even more to smallpox.

I recall returning soaking wet, cold, and miserable, more than half hypothermic after a backpacking trip in the gloriously natural Canadian Rockies, during which fog and mist had alternated with rain, hail, and snow (in August, mind you!), and then encountering this bit of wisdom from 19th-century English writer and art critic, John Ruskin: "There is no such thing as bad weather, only different kinds of good weather"—whereupon I concluded that Mr. Ruskin hadn't spent much time in the mountains. Similarly, I suspect that those well-intentioned people who admire "natural" raw milk have never experienced the ravages of *Campylobacter*, pathogenic *E. coli*, or bovine tuberculosis, each spread by unpasteurized milk.

Even in sports, with its cult of the "natural" athlete, devotees strive to move beyond the natural to what is beautiful, elegant, or impressive, fully recognizing that it takes work and practice. Hence there is spring training and interminable "practice" as well as exhibition games, coaches, and trainers. Dressage (a classical form of horsemanship) seeks to help a horse move with a mounted rider as beautifully as it would solo, in nature. To do this takes at least a decade of effort, pushing horse and rider to work hard and in *unnatural* ways to ultimately achieve harmony and beauty. It is natural for horses to stand around in fields, eating and pooping and swatting flies with their tails. It is not natural for a horse to dance to music. Training the horse brings out its natural beauty, but only after enormous amounts of "unnatural" labor. Similarly, it is not natural to climb high mountains. Mountaineering is difficult and stressful, requiring physical conditioning, skill, and technical equipment. But I am not alone in my sense that getting to the top of a mountain is a "peak experience" indeed, worth the time, trouble, and risk.

In short, natural is often good, but not always. It may be natural to be a couch potato, to punch someone in the nose if he has angered you, for people to get sick, or for a child to resist toilet-training. And of course, bacterial infections, lousy weather, and troublesome behavioral inclinations aren't the only regrettable entities out there in the natural world. Don't forget hurricanes, tsunamis, earthquakes, droughts and the devastation wrought by volcanoes, lightning storms, sandstorms, blizzards, and the combustion of entirely natural fossil fuels.

In his book, *A Treatise on Human Nature*, published 1739/1740, David Hume presented, and criticized, what has come to be known as the "is-ought problem," the notion that we can derive what ought to be from an examination of what is. Is there any way, Hume asked, that we can legitimately connect how the world "is" (which by extension includes our own behavioral inclinations) with how it "ought" to be (including how we ought to behave)? At least one respected modern philosopher, Max Black, felt that simply by raising the question, Hume so conclusively severed "is" from "ought" that he called this distinction—between the *descriptive* and the *prescriptive*, or between *facts* and *values*—"Hume's Guillotine." Hume's insight, that it is fallacious to derive "ought" from "is," has also come to be known as the "naturalistic fallacy," a term coined by English philosopher G. E. Moore in his 1903 book, *Principia Ethica*.

In 1710, three decades before Hume sliced into the is-ought problem, philosopher and mathematician Gottfried Leibniz had struggled with the problem of "theodicy," the theological effort of reconciling the existence of evil and suffering in a world supposedly governed by a god that is both all-powerful and wholly benevolent. It was—and still is—a tall order. Leibniz concluded that because God is necessarily good (by Judeo-Christian-Islamic definition, at least), as well as omnipotent, and because this deity evidently chose to make the world as it is, in view of all the possible ways that it might have been, then this must be "the best of all possible worlds" (*le meilleur des mondes possibles*). This famous phrase has proven easy to satirize, most notably by Voltaire in his novel *Candide*, the picaresque adventures of Dr. Pangloss (Voltaire's Leibniz caricature) and his student Candide, who experience no end of terrible events—always interpreting them through a ridiculously cheerful, positive lens.

Voltaire was especially outraged by a particularly devastating natural disaster, the Lisbon earthquake of 1755, which is estimated to have killed tens of thousands of innocent people. But he also wasn't shy about depicting the cruel but equally "natural" behavior of murderers, rapists, and torturers. It's a theme that continued to resonate.

In the 19th century, John Stuart Mill argued in his essay *Nature* that "nature cannot be a proper model for us to imitate. Either it is right that we should kill because nature kills; torture because nature tortures; ruin and devastate because nature does the like; or we ought not to consider what nature does, but what it is good to do."

By the same token, it would be a misuse of this book for anyone to conclude that polygamy, because it is "natural," is good.

Modern evolutionary biology makes it clear that "nature," acting through natural selection, whispers in our ears—cajoling, seducing, imploring, sometimes even threatening or demanding—and undoubtedly inclining us in one direction or another. These inclinations, in turn, are derived from a remarkably simple process: the automatic reward that comes from biological success. If a given behavior leads to greater eventual reproductive success on the part of the "behaver" (more crucially, heightened success for any genes that predispose toward the act in question), then selection will promote those genes, and thus, the behavior. It will seem—and be—natural. And although "good" for the future of the relevant genes, this is a far cry from anything remotely approximating moral or ethical desirability.

Natural selection has a very efficient way of getting animals and people to do things that are "good" for the organism—or at least that contribute to the success of those genes that generate a propensity for doing those things: call it pleasure. Living things find it pleasurable to eat when hungry, drink when thirsty, sleep when tired, obtain sexual satisfaction when aroused, and so forth. Genes that induced their bodies to, say, refrain from nourishing themselves would not be well served over evolutionary time. And so, eating when hungry or drinking when thirsty "feel good" because feeling good is what gets animals and people to do those things, and organisms that didn't make the "do such-and-such/feel good" connection would have left fewer descendants than those for whom pleasure basically tracked reproductive success, what evolutionists call "fitness." But whether, in Mill's terms, such things are necessarily "good to do," in the sense of ethics and morality, is another matter entirely.

Gravity exists, quite naturally. But few people would derive ethical guidance from the natural world of physics, which would mean no more standing upright because this goes against such a universal precept. Should we refrain from cleaning the house because the Second Law of Thermodynamics—another fundamental, natural law—dictates that disorder necessarily increases within any closed system and therefore we must conclude that entropy is good and struggling against it is wrong? Is it unethical to exceed the speed of light, or simply impossible? Similar absurdities arise if one attempts to "naturalize" ethics from chemistry, geology, astronomy, mathematics, and so forth. When it comes to evolutionary biology, however, many people seem to feel otherwise.

After all, isn't there something good—maybe even magnificent—about brilliant autumn foliage, the song of a nightingale, the majesty of a bull elephant? If nothing else, they bring pleasure, even delight, to people. And isn't it downright *good* for a mother robin to feed her nestlings? Doing so is certainly good for the baby robins, and thus for the evolutionary success of the adults—more precisely, for any genes that predispose adult bird bodies to feed their offspring. Leaving aside, for the moment, the less-than-salubrious effect this has on those worms whose lives are thereby cut short, it is easy to assume that the working of biological nature—as distinct, perhaps, from physical nature, or chemical nature, or geological nature—is not only admirable at the level of observing human intellects but also ethically instructive.

But a dispassionate look at the natural world seems to lead, if anything, to confirmation that the naturalistic fallacy is indeed fallacious. Not that it denies basic logic to say that if something is biologically natural it must be good; this isn't equivalent to concluding, say, that if Socrates is a man and all men are mortal, then Socrates is *not* mortal. Rather, it simply does not follow that biological nature is *necessarily* good, in the sense of giving us insight into morality or ethics. To an extent that should trouble any "natural ethicist," the living world is a zero-sum game in which benefit for one organism often comes at the expense of others, and where no sign of overarching ethical restraint, no independent claim to goodness, can be discerned.

Annie Dillard's marvelous meditation on nature, *Pilgrim at Tinker Creek,* contains an unforgettable account of her encounter with a "very small frog with wide, dull eyes." As Dillard describes it

> Just as I looked at him, he slowly crumpled and began to sag. The spirit vanished from his eyes as if snuffed. His skin emptied and drooped; his very skull seemed to collapse and settle like a kicked tent. He was shrinking before my eyes like a deflating football. I watched the taut, glistening skin on his shoulders ruck, and rumple, and fall. Soon, part of his skin, formless as a pricked balloon, lay in floating folds like bright scum on top of the water: it was a monstrous and terrifying thing. I gaped bewildered, appalled. An oval shadow hung in the water behind the drained frog; then the shadow glided away.

This "shadow," which had just killed the frog—so smoothly, mercilessly, and naturally, before gliding away—was an enormous, heavy-bodied brown creature that (to quote Dillard, once again)

> eats insects, tadpoles, fish, and frogs. Its grasping forelegs are mighty and hooked inward. It seizes a victim with these legs, hugs it tight and paralyzes it with enzymes injected during a vicious bite. . . . Through the puncture shoot the poisons that dissolve the victim's body, reduced to a juice. This event is quite common in warm fresh water. The frog I saw was being sucked by a giant water bug. . . . I stood up and brushed the knees of my pants. I couldn't catch my breath.

Dillard quickly goes on, achieving a more objective and scientific detachment:

> Of course, many carnivorous animals devour their prey alive. The usual method seems to be to subdue the victim by downing or grasping it so it can't flee, then eating it whole or in a series of bloody bites. Frogs eat everything

whole, stuffing prey into their mouths with their thumbs. People have seen frogs with their wide jaws so full of live dragonflies they couldn't close them. Ants don't even have to catch their prey: in the spring they swarm over newly hatched, featherless birds in the nest and eat them tiny bite by bite.

Dillard notes, with understatement, that "it's rough out there, and chancy," that "every live thing is a survivor on a kind of extended emergency bivouac," and that "cruelty is a mystery," along with "the waste of pain." And finally, she concludes that we "must somehow take a wider view, look at the whole landscape, really see it, and describe what's going on here. Then we can at least wail the right question into the swaddling band of darkness, or, if it comes to that, choir the proper praise."

In Stephen Sondheim's dark musical, *Sweeney Todd*, we learn that "the story of the world, my sweet, is who gets eaten and who gets to eat." In the nonfiction world we all inhabit, there is pleasure and pain; suffering and delight; eaten and eater; life and death; growth and decay; luscious Louisiana salt-marshes teeming with innumerable, natural, glorious lives and horrendous pollution and destruction wrought in 2010 by noxious eruptions of wholly natural petroleum.[10] One could argue that nothing on this planet can ever be unnatural because everything is necessarily composed of the elements of the periodic table, all of which are "natural." And yet, just a moment's reflection—or, better yet, the instantaneous insight of nonreflection—tells us that some things had best be left alone, Pandora's boxes that should never be opened, and, if folly and avarice does so nonetheless, that these things must be struggled against with all the strength and determination, natural or not, that we possess.

In his *Essay on Man*, Alexander Pope seconded Leibnitz and gave poetic voice to a widespread perception:

> All nature is but art, unknown to thee;
> All chance, direction which thou canst not see;
> All discord, harmony not understood;
> All partial evil, universal good;
> And spite of pride, in erring reason's spite,
> One truth is clear: whatever is is right.

I strenuously disagree.

Evolution by natural selection is an extraordinarily important and endlessly fascinating subject. It has produced you and me and every other living creature. But good it isn't! My first point, accordingly, is that natural selection is every bit as natural as Newton's Laws, Planck's constant, or relativity, and every bit as devoid of moral direction. Like the laws of physics, the laws of biology simply describe what is, not what should be.

My second point is that if anything, the evolutionary process is more negative than neutral when it comes to humane values; it is likely to lead to results that most ethicists will, and should, reject. Note: I am not counseling a rejection of natural selection or of evolution as a historical, biological, and natural process; rather, I reject the intimation that natural selection is somehow good and that we ought to align our lives to conform to its specifications.

As biologist George C. Williams pointed out, our current understanding of natural selection is that it operates as a ratio, with the numerator reflecting the success of genes in projecting copies of themselves into the future and the denominator, the success of alternative genes. Because a gene (or an individual, a population, even—in theory—a species) maximizes its success by producing the largest such ratio, it can do so either by reducing the denominator or increasing the numerator. Most creatures, most of the time, find it easier to do the latter than the former, which is why living things generally are more concerned with feathering their nests than de-feathering those of others.

Taken by itself, such self-regard isn't the stuff to make an ethicist's heart go pitter-pat. But to make matters worse, animal studies in recent years have revealed a vast panoply of behavior whereby living things have no hesitation in minimizing the denominator, trampling over others in pursuit of their own biological benefit. We have long known that the natural world is replete with grisly cases of predation, parasitism, a universe of horrors all generated by natural selection and unleavened by the slightest ethical qualms on the part of perpetrators.

Worse yet, perhaps, are the cases of vicious genetic self-promotion at the expense of others within the same species. Earlier we reviewed the reality of infanticide, now identified in numerous species—notably those prone to polygyny, including lions and many nonhuman primates such as langur monkeys and chimpanzees. This phenomenon, natural as can be, is so horrifying that even hard-eyed biologists had a difficult time accepting its ubiquity, and—until recently—its "naturalness" as well. But natural it is, and a readily understood consequence of evolution as a mindless, automatic, and value-free process whose driving principle is not just amoral but—by any decent human standard—downright immoral.

"Natural selection," wrote George C. Williams,[11] one of the giants of 20th century evolutionary theory, "can honestly be described as a process for the maximization of short-sighted selfishness." Add cases of animal rape, deception, nepotism, siblicide, matricide, and cannibalism, and it should be clear that natural selection has blindly, mechanically, yet effectively favored self-betterment and self-promotion, unmitigated by any ethical considerations. I say this fully aware of an important recent trend in animal behavior research: the demonstration that animals often reconcile, make peace, and cooperate. The sobering reality, however, is that no less than the morally repulsive examples just cited, these behaviors *also* reflect the profound self-centeredness of the evolutionary process—increasing the numerator of the "fitness ratio" by enhancing either one's own success or that of one's genetic relatives. If the outcome in certain cases is less reprehensible than outright slaughter, it is only because natural selection only sometimes works to reduce the denominator. But all the time, the only outcome assessed by natural selection is whether a given tactic works—whether it enhances fitness—not whether it is good, right, just, admirable, or in any sense moral.

Why, then, should we look to such a process for moral guidance? Instead, insofar as evolution has engendered behavioral tendencies within ourselves that are callously indifferent to anything but self- (and gene-) betterment, and armed as we now are with insight into the origin of such tendencies and—crucially important, the ability to rise above them if we choose—wouldn't sound moral guidance suggest that we intentionally act contrary to them?

In the movie *The African Queen*, Katherine Hepburn's character stiffly observes to Humphrey Bogart, "Nature, Mr. Allnut, is what we were put on earth to rise above." I strongly doubt that we were put on earth to do anything in particular, but if we want to be ethical—rather than simply "successful"—rising above our human nature may be just what is needed. Evolution by natural selection, in short, is a wonderful thing to learn *about*, but a terrible thing to learn *from*.

By the end of the 19th century, Thomas Huxley was perhaps the most famous living biologist, renowned in the English-speaking world as "Darwin's bulldog" for his fierce and determined defense of natural selection. But he defended evolution as a scientific explanation, not as a moral touchstone. In 1893, Huxley made this especially clear in a lecture titled "Evolution and Ethics," delivered to a packed house at Oxford University. "The practice of that which is ethically best," he stated,

> what we call goodness or virtue—involves a course of conduct which, in all respects, is opposed to that which leads to success in the cosmic struggle for existence. In place of ruthless self-assertion it demands self-restraint; in place of thrusting aside, or treading down, all competitors, it requires that the individual shall not merely respect, but shall help his fellows; its influence is directed, not so much to the survival of the fittest, as to the fitting of as many as possible to survive.

"The ethical progress of society depends," according to Huxley, "not on imitating the cosmic process [that is, evolution by natural selection], still less in running away from it, but in combating it."

It may seem impossible for human beings to "combat" evolution because *Homo sapiens*—just like every species—is one of its products. And yet, Huxley's exhortation is not unrealistic. It seems likely, for example, that to some extent each of us undergoes a trajectory of decreasing selfishness and increasing altruism as we grow up, beginning with the infantile conviction that the world exists solely for our personal gratification and then, over time, experiencing the mellowing of increased wisdom and perspective as we become aware of the other lives around us, which are not all oriented toward ourselves. In her novel *Middlemarch*, George Eliot noted that "we are all born in moral stupidity, taking the world as an udder with which to feed ourselves." Over time, this "moral stupidity" is replaced—in varying degrees—with ethical acuity, the sharpness of which can largely be judged by the amount of unselfish altruism that is generated.

Many sober, highly intelligent scientists and humanists nonetheless misunderstand the connection between evolution and morality, grimly determined that evolutionary facts are dangerous because they justify human misbehavior. The renowned developmental psychologist Jerome Kagan exemplifies this blind spot. "Evolutionary arguments," he writes, "are used to cleanse greed, promiscuity, and the abuse of stepchildren of moral taint."[12] Similar arguments were in fact used in this way, in the unlamented days of social Darwinism. But no longer.

Moreover, an argument can now be made that certain cultural practices—such as those encouraging monogamy, or at least discouraging the excesses of polygamy—may help to cleanse our ancestral, polygamous tortoise of some of its troublesome inclinations. Ape-men and ape-women, we are also the only earthly organisms capable of assessing our own nature and then deciding what—if anything—to do about it.

Notes

1. Barash, D. P. (1987). *The hare and the tortoise: Culture, biology and human nature.* New York: Penguin.
2. de Beauvoir, S. (1961). *The second sex.* New York: Bantam Books.
3. Dickemann, M. (1981). Paternal confidence and dowry competition: A biocultural analysis of purdah. In R. D. Alexander & D. W. Tinkle (Eds.), *Natural selection and social behavior.* New York: Chiron Press.
4. Similarly, although it appears that women's breasts are to some extent eroticized across nearly all human cultures, it is well-known that this is especially true in societies in which breasts are typically obscured, compared to certain tropical societies in which bare breasts are normal and thus less sexually charged.
5. Cooper, E. (1917). *The harem and the purdah: Studies of Oriental women.* New York: Century.
6. King, M. L., Jr. (2002). *A call to conscience: The landmark speeches of Dr. Martin Luther King, Jr.* New York: Grand Central Publishing.
7. Collard, M. (2002). Grades and transitions in human evolution. In T. J. Crow (Ed.), *Proceedings of the British Academy: Vol. 106. The speciation of modern homo sapiens.* Oxford, England: Oxford University Press.
8. Rice, W. R. (2000). Dangerous liaisons. *Proceedings of the National Academy of Sciences, 97,* 12953–12955.
9. Arouet de Voltaire, F. M. (2009). *The works of Voltaire: A contemporary version with notes.* Ithaca, NY: Cornell University Press.
10. Oil is largely the result of huge quantities of oceanic plankton (mostly diatoms), compressed along with terrestrial vegetation and marinated underground for millions of years: entirely natural and yet grotesquely unpleasant.
11. Williams, G. C. (1988). Huxley's *Evolution and Ethics* in sociobiological perspective. *Zygon, 4,* 385–400.
12. Clark, M. E. (2002). *In search of human nature.* London: Routledge.

9

Losing Illusions

Evolutionary biology is *the* foundational scientific understanding of how living things came to be as they are, and hence, what they are all about. As philosopher Daniel Dennett put it, evolution by natural selection is the "best idea anyone ever had."[1] For many, it is also among the most difficult. Not that it is hard to understand, although it is trickier than it seems. To revise Winston Churchill's encomium to the Royal Air Force during the Battle of Britain, never have so many written and said so much about something they understood so poorly. What is especially challenging, even beyond understanding its subtleties—such as the fact that it maximizes individual genetic success and doesn't work for the "good of the species"—is the extent to which evolutionary biology forces us to rethink some of our most cherished illusions.

Most prominently, evolution requires us to get over the illusion of unique human centrality. Many people still hold to the comforting myth that our species was specially created, by God and in His image. Even among those not benighted by religious fundamentalism, it is common for people to retain the notion that we are still, in some inchoate way, SPECIAL—if not the apple of God's eye, at least biologically distinct from the rest of life.

This is a version of what might be called Brahe's Blunder, after the brilliant 16th-century astronomer Tycho Brahe, whose calculations showed that the known planets of his day—Mercury, Venus, Mars, Jupiter, and Saturn—circled the sun. Forced to accept these empirical verities, yet deeply committed to a Biblical worldview whereby the Earth had to be the center of the universe, Tycho Brahe shoehorned his findings into a more acceptable system: he proposed that the sun with its five-planet retinue nonetheless revolved around a central and immobile Earth! I propose, in turn, that many people commit a modern version of Brahe's Blunder, accepting evolution by natural selection as scientific fact (which indeed it is), yet in their heart of hearts retaining the illusion that we aren't "really" animals like all those other creatures.

But we are. The most resonant take-home message from evolution is one of continuity among all living things, reflected not only in shared historical connectedness via the fossil record but also via a fundamental coherence that links all of life. To be sure, this "sameness" is shot through with numerous distinctions; this, after all, is the basis for identifying *any* species—including our own—as different from others.[2]

But no matter how troublesome it may be to well-meaning persons, the reality is that there is no qualitative discontinuity between our biology and that of other organisms. Understanding our polygamous nature without denying ourselves the prospect of self-determination is but one of many stopping places as we outgrow fairy-tale illusions of cosmic centrality and spiritual specialness, along our journey to deeper self-knowledge.

"Nothing is easier," wrote Darwin,

> than to admit in words the truth of the universal struggle for life, or more difficult—at least I have found it so—than constantly to bear this struggle in mind. Yet unless it be thoroughly engrained in the mind, the whole economy of nature, with every fact on distribution, rarity, abundance, extinctions and variation, will be dimly seen or quite misunderstood.

The preceding appeared in *The Origin of Species*, which notably avoided any discussion of human beings. Here, as in so many other places, Darwin's observation is acute, even if restricted to the nonhuman organic world. But it is no less true of *Homo sapiens*. I am reminded of Alexander Pope's epitaph for Isaac Newton: "Nature and Nature's laws lay hid in night: God said, Let Newton be! and all was light." Without evolution by natural selection, much about ourselves lies hidden, and we, benighted. With it, the scales fall from our eyes.

At the same time, although truths about our own nature may indeed set us free, it is nowhere promised that these same truths will be painless or easy to swallow. We are, after all, conflicted creatures, not least when it comes to the most intimate details of our lives. Earlier, I quoted from Pope's *Essay on Man*, seconding the poet's awareness of our confused and confounded nature. In that same work, just after famously announcing that "the proper study of mankind is man," Pope described our species as having been

> Plac'd on this isthmus of a middle state,
> A being darkly wise, and rudely great:
> With too much knowledge for the sceptic side,
> Alike in ignorance, his reason such,
> Whether he thinks too little, or too much:
> Chaos of thought and passion, all confus'd;
> Still by himself abus'd, or disabus'd

The poet was meditating in particular on the dichotomous tension between human reason and passion:

> Man's superior part
> Uncheck'd may rise, and climb from art to art;
> But when his own great work is but begun,
> What Reason weaves, by Passion is undone.

He might as well have been bemoaning the simultaneous appeal of pair-bonding and monogamy on one hand and polygamy on the other. We are complex critters, endowed with many conflicting inclinations, whereby "Reason" (plus love) calls for marital fidelity, while "Passion" (plus temptation) whispers in our other ear.

Evolution is a powerful predictor of these predilections, a useful guide to why we do what we do. It is a unique instrument of intellectual optics, both a magnifying lens, enabling us to see tiny inclinations up close, and a telescope, helping us spot trends that otherwise are impossibly distant. "We have evolved a nervous system that acts in the interests of our gonads," writes Michael Ghiselin,[3] "and one attuned to the demands of reproductive competition. If fools are more prolific than wise men, then to that degree folly will be favored by selection. And if ignorance aids in obtaining a mate, then men and women will tend to be ignorant."

This is true enough, and regrettably so. The same holds for violence (under some circumstances), sexual voraciousness (ditto), intolerance of the escapades of one's mate while being disposed to engage in comparable behavior oneself, and so on. At the same time, ignorance as such would seem ill-advised. When it comes to our own proclivities as well as the predispositions of others that we care about—or whose behavior impacts our selves and our loved ones—we are well advised to replace ignorance with wisdom, or at least, with insight into what lies deep within our shared human nature.

But even as we embrace biological explanations of human behavior, there is a danger of doing so with too much enthusiasm, thereby foreclosing our very real human potential for adjustment and change. Despite our shared polygamous history and the fact that somewhere deep inside everyone there lurks the ghosts of our harem-holding (for men) and harem-hiding (for women) pasts, by virtue of being human we also have the capacity—should we choose to do so—to make conscious, deliberate, mindful choices that refute our polygamously generated inclinations. These inclinations whisper within us, but they do not shout; they generate tendencies and influences, not rigid requirements.

Perhaps the most notable characteristic of *Homo sapiens* is that even though we are no less "biological" than other organisms, we possess the unique ability to define ourselves by how we choose to live, by deciding which inner whisperings we elect to follow and which to suppress and/or redirect. In short, in this chapter I will make the case that we can defy our polygamous heritage if (1) we understand it well enough to comprehend its influence over us and (2) we make the informed decision to do so.

In *Middlemarch*, George Eliot wrote that "there is no creature whose inward being is so strong that it is not greatly determined by what lies outside it." True enough for human beings as well, although in our case, it would be appropriate to substitute "influenced" for "greatly determined." In chapter 4, we noted that Gunnison's prairie dogs, among other creatures, vary their mating system between polygyny and monogamy depending on the patchiness of available resources. If they can do it, why can't we? To be sure, we do, and not simply as a consequence of resource distribution. A mere glance at human diversity shows that we aren't irrevocably stuck into any one way of living.

It is easy to misunderstand exactly what biology contributes to complex human behavior, partly because that contribution is often murky. But that doesn't mean it isn't real. Misunderstandings especially arise when it is assumed that biology generates absolutes, whereas in fact—and notably in the case of polygamy versus monogamy—it often results in predispositions rather than rigid determinism. Stephanie Coontz, perhaps our preeminent historian and sociologist of modern marriage, fell into this error when she wrote the following about the diversity of human mating systems:

> The rules governing who can marry whom have varied immensely from group to group. In some societies, marriages between first cousins have been prohibited. In others, such unions have been preferred. Some societies have encouraged polygamy. Others strictly prohibit it. Such a contradictory hodgepodge of social rules could not have sprung from some universal biological imperative.[4]

They indeed could not have. But a biological inclination—different from a "universal biological imperative"—is totally consistent with such a diversity, or "hodgepodge." It is especially consistent with a strong statistical bias, as we find in the case of polygamy, just as it fits impressively with the array of other traits, anatomical as well as behavioral, that we have already considered. Human beings, to be sure, are unique among other animals in that we often negotiate social and sexual relationships rather than acting according to a predetermined script. Female bonobos, for all their social power, do not evaluate their daughters' potential boyfriends, urging one over another. Nor do chimpanzees—a species in which females typically emigrate from their birth troop and end up mating within a different one—orchestrate daughter exchanges to maximize their own social capital.

When it comes to illusions that we need to lose, a prime candidate is the false notion that biology determines our behavior, including complex social and sexual patterns. What it does, however, is *influence* them, with that influence readily apparent not only in its breadth—the extent to which such patterns play out cross-culturally—but also in its depth: the degree to which it impacts each human being, regardless of his or her social circumstance, cultural setting, or even personal preference.

In this regard, there is much to be said for understanding our nonhuman primate relatives. Research can provide tremendous insights into general principles that apply to our own species no less than to others: bimaturism, sexual dimorphism in behavior as well as structure, and so forth. But skepticism is needed when it comes to naïve extrapolations from one species to another. Thus, human beings partake of male–male bonding (as in chimps), female–female bonding (as in bonobos), and also a high frequency of nuclear families (found in neither ape). We're not as polygynous as gorillas, not as monogamous as gibbons, and definitely not predisposed to the make-love-not-war, pan-erotic hijinks of bonobos or the more aggressive sexual free-for-all of chimps. Nor do we engage in the expanded polygyny of orangutans, in which the territory of isolated, mutually antagonistic males typically incorporates the ranges of several females. Although we have more paternal involvement than any other primate, we are definitely less paternal than maternal. We are ourselves—easily said, but not nearly so easy to pin down.

How many different ways are there of being human? If we had it to do over again, and then again and yet again, how many different lives could we lead? A hundred? A million? Very many, to be sure, but the possibilities are not infinite. The great variety of human cultural traditions is impressive, so much so that cultural anthropologists have been kept busy for decades merely cataloging the variations. The point of this book, however, has been that such variations follow a limited number of themes. Just as a family going to pick out a Christmas tree, for example, will focus on the differences between them (this one is taller, that one has more branches or is shaped more symmetrically), observers of human behavior have tended to focus on the differences between, say, the lifestyles of a New Guinea highlander and a New York taxi driver. But an objective Martian scientist, considering a Christmas tree farm for the first time, would almost certainly be struck, right off, not by the differences among the trees but by their similarity. By the same token, an objective, biologically based view of human social arrangements cannot help but notice the fundamental *similarities*—including those that speak to a polygamous commonality—that underlie the obvious cultural variations.

In short, there is such a thing as human nature, just as there is hyena nature, halibut nature, and, for that matter, hickory tree nature. Consider the simple fact that our species has often been at odds with its social environment—notably, with enforced monogamy, but even more with mandated celibacy. Most of the world's great social philosophers have described situations in which people and their societies are in conflict. If we were simply the passive products of our social upbringing, such conflict would not occur; we would be rubbed smooth and rendered altogether accommodating by our experiences. But no: we are restless and cantankerous creatures, rough and prickly, or—to switch the metaphor—often like square pegs shoved into round holes. If, instead, we were formed of modeling clay, if we were putty to be molded passively by our environment, then human beings and their societies would not conflict.

I also think it unlikely that most men consciously intend to oppress women, although I freely acknowledge that women have long been oppressed. An understanding of sex differences leads to the conclusion, for example, that men are likely to employ an aggressive style, an evolutionary bequeathal largely due to the payoffs associated with prehistoric success in male–male competition. Possessing this "style," and to some degree possessed by it, it seems that men are rather pushy creatures whose aggressive tendencies roll like a steamroller over whatever is in their path: other men, women, puppy dogs, rain forests, and so forth. When it comes to the oppression of women, there is undoubtedly some misogyny involved, at least in specific cases; but most of the time, men are less concerned with oppressing women than with oppressing, suppressing, or repressing *anyone* who gets in their way.

French philosopher, historian, and social critic Michel Foucault began his oft-confusing work *The History of Sexuality* by claiming that the Western world labored under a "repressive hypothesis" whereby from the 17th to the mid-20th century, indications of sexuality in daily life were repressed, whereas in fact—he argued—various departures from traditional marital sex were reported, examined, and highly influential. I believe, au contraire, that this repressive hypothesis was more accurate than Foucault realized. To be sure, there were raunchy novels, dirty jokes and poems, mistresses and extra-marital lovers, as well as houses of prostitution. And now, in the 21st-century West, sex is ubiquitous: on the Internet, advertising, in our faces no less than in our overheated imaginations.

And yet, when it comes to the most widespread assumptions beneath the wildly gyrating sexual kaleidoscope of modern life, most people still retain something very close to the "repressive hypothesis" and, like the French police chief in the movie *Casablanca*, are liable to be "shocked, shocked" to learn that monogamy isn't the "natural" human condition. Many in the supposedly open-minded West continue to believe, with Foucault, that our shared human condition is one of "the monotonous nights of the Victorian bourgeoisie" in which sexuality is "carefully confined" and "moved into the home." Here is Foucault, in more detail:

> The conjugal family took custody of it [sex] and absorbed in into the serious function of reproduction. On the subject of sex, silence became the rule. The legitimate and procreative couple laid down the law. The couple imposed itself as model, enforced the norm, safeguarded the truth, and reserved the right to speak while retaining the principle of secrecy. A single locus of sexuality was acknowledged in social space as well as at the heart of every household, but it was a utilitarian and fertile one: the parents' bedroom. The rest had only to remain vague; proper demeanor avoided contact with other bodies, and verbal decency sanitized one's speech. And sterile behavior carried the taint of abnormality; if it insisted on making itself too visible, it would be designated accordingly and would have to pay the penalty.[5]

By "sterile behavior," Foucault meant homosexuality, as well as childhood sexuality plus any number of "perversions" that powerful, polite society considered inappropriate, subversive, and thus dangerous. He wrote very little about polygamy as such. The underlying reality of human polygamous tendencies—which, ironically, are motivated by evolution and thus far from "sterile"—are more widespread, as well as more socially disruptive, than homosexuality, childhood sexuality, and anything else that Foucault could imagine (or engage in), combined. And his imagination, as well as his life, was pretty technicolor.

Turning now specifically to polygamy, there are two particular illusions that deserve to be dispelled, one that overestimates the impact of our polygamous evolutionary history and another that underestimates it. If you are firing at a target and your first shot is off 40 degrees to the left, and your second, 40 degrees to the right, then you have hit the bullseye . . . statistically speaking. In reality? Not so much. When it comes to humanity and polygamy, reality splits the difference, quite precisely, between two equally erroneous and extreme views.

First, the overestimate is the claim that polygamy is so natural and normal, and therefore healthy and appropriate, that monogamy is an evil, artificial institution by which society inhibits the otherwise beneficent emergence of our pan-sexual selves. This view is especially popular in the polyamorous community, many of whom consistently underplay the strong undertow of sexual jealousy, not realizing that jealousy is every bit as natural and normal as polygamy itself, and equally difficult to abolish by ideological fiat. The most recent popular manifestation of this illusion can be found in the book *Sex at Dawn*,[6] a truly egregious misrepresentation of biological and anthropological fact, gulling the reading public in the service of maximizing sales as well as the authors' evident desire to justify their own chosen lifestyle.

Sex at Dawn is so grotesquely flawed that it warrants an entire book to refute its numerous misunderstandings of evolution as well as its blatant factual misrepresentations of primatology, psychology, and anthropology—in short, the extent to which it misleads nonscientist readers. Fortunately, such a book has already been written: *Sex at Dusk*, by Lynn Saxon.[7] In it, Saxon notes that *Dawn* "constantly reminded me of a line from the novel *Nice Work* by David Lodge (1988): 'Literature is mostly about having sex and not much about having children; life's the other way round.'" Ms. Saxon goes on to observe that *Sex at Dawn* is "almost all about sex and not much about children, yet evolution is very much about reproduction—variation in reproductive success is evolution."

Sex at Dawn gives notably short shrift—indeed, no "shrift" at all—to paternal love; this isn't surprising because under the fictitious sexual regime that it postulates, there would be little or no paternal love and affiliation. But paternal love and devotion are real, precisely because there is sexual jealousy, which is selected for as a means of enhancing confidence of genetic relatedness in a species whose slow-developing offspring need all the adult care and attention they can get. In highly polygynous species, there is little time or opportunity for the harem-holding male to invest in his individual children, and in promiscuity (because of the lack of paternal confidence), there is even less reason for males to do so. By contrast, in monogamy, there is both; indeed, as we have seen, this seems to be a main reason—probably *the* main reason— for the evolution of monogamy itself, whenever it occurs.

Sex at Dawn is single-mindedly obsessed with making a case that human beings are naturally polyamorous. Toward this end, much of the so-called "group sex" it reports is actually gang rape, used both to subjugate women and to reduce sexual tension and competition among men. Its claims about communal sex among prehumans are nothing but armchair anthropology and wishful thinking, enlivened by prurience in the service of an agenda that, quite simply, doesn't hold water and is consistently refuted by the details of human genital and spermatic anatomy, among other sources. After an exhausting review of primate (including human) reproductive anatomy and physiology, primatologist Alan F. Dixson concluded that "it is highly unlikely that *Homo sapiens* is descended directly from an ancestor which had a multi-male/multi-female mating system [i.e., the kind of polyamorous, promiscuous system promised by the authors of *Sex at Dawn*]. Descent from a polygynous or monogamous precursor is much more likely."[8]

Sex at Dawn makes much of the sexual practices of one group of people, known as the Mosuo (sometimes called the "Na") of China. These people indeed appear to be notably "liberated," especially their women: more so, perhaps, than any other human social group.[9] However, their uniqueness—notably their custom of "walking marriages" whereby a man can spend the night in the sleeping quarters of a woman, with very little guilt or obligations affixed to either "husband" or "wife"—is actually a cogent argument *against* the claim that human beings are inherently predisposed toward guilt-free sexual laxity, because the Mosuo are highly unusual and definitely not representative of human beings generally.

For a somewhat analogous situation, the Yanomamo of the Venezuelan Amazon are renowned as the "fierce people," and they really are unusually violence prone. But it is misleading to generalize from the Yanomamo to "human nature" as a whole when it comes to violence in general and primitive "war" in particular, because there are many other traditional human groups that are much more pacific. By the same token, it is misleading to generalize from the Mosuo to human beings as a species. In fact, extrapolating from the Mosuo in this respect is much less justified than in the case of violence and the Yanomamo because although there are other human groups—in addition to the Yanomamo—that are fierce and violence prone, there are *no other people* whose sexual practices are comparable to the Mosuo. Taking this single example, what scientists call an "*n of 1*"—even though in this case it refers to a group—and claiming that it points the way to genuine human nature is like taking the height of the Boston Celtics starting basketball team and claiming that this demonstrates the height of the average human being.

In their search for anything that supports their argument, the authors of *Sex at Dawn* attribute great import, as well, to the exaggerated sexuality of bonobos, in the course of which the reality of bonobo sexuality is itself exaggerated as well as erroneously presented. Thus, among bonobos, sex is overwhelmingly used to deflect aggression and to establish and reinforce dominance, not to achieve social bonding via a pan-sexual utopia.[10] Moreover, bonobos are no more closely related to us—genetically and thus, evolutionarily—than are chimpanzees, which are considerably more prone to sexual jealousy.

The evidence is overwhelming that in no human society is sexual promiscuity considered the norm, and, moreover, we have already observed that basic anatomical evidence shows convincingly that we did not evolve in a multimale, multifemale (polygynandrous) sexual cornucopia. We are not bonobos, never have been, and almost certainly never will be. To paraphrase Dorothy Parker, books that ignore science and replace it with such *Dreck at Dawn* shouldn't be tossed aside lightly: they should be thrown away with great force.

Now for the other extreme: the equally misleading illusion that monogamy is not only desirable (which it may well be) but also natural, which it assuredly is not. There are several dangers to this viewpoint. For one, in underestimating the potency of polygyny and polyandry, it leaves its followers vulnerable to being blindsided when polygamy rears its multifaceted head, whether within themselves or among those they care about, notably their "significant others." In proportion as someone buys into the myth that monogamy is normal and natural, then departures from monogamy—whether in actuality or merely "in one's heart"—are nasty surprises indeed. Better to be prepared, to look both ways, and to know that polygamous inclinations, at least as much as monogamous ones, are normal and natural.

This doesn't mean that such inclinations need to be followed, or should be; that is a matter for each individual to resolve. But it does mean that everyone should be aware of the prospect—indeed, in most cases, the certainty—of sexual temptation, just as we should be aware of the temptation exerted by rich foods or the prospect of cheating on our income tax. Moreover, the fact that such temptations arise does not mean that a person is weak, or sinful, or mated with the wrong person; rather, it simply means that he or she is a healthy primate!

Let's be clear: monogamy isn't natural, but as emphasized in the previous chapter, some of the best things we do aren't those that "come naturally" but that involve hard work, often going specifically against our inclinations. It's easier to be lazy than to do good work. Easier not to learn a second language, not to practice playing the violin, to be a couch potato instead of an athlete. It is easy to do what comes naturally: to breathe when our CO_2 levels reach a certain level, to sleep when tired, eat when hungry, drink when thirsty, and yes, have sex when horny. None of these things, however, are uniquely human. All are deeply biological and none especially admirable. To be maximally useful, "sex ed" should therefore go beyond merely the birds and bees, and the details of male–female plumbing; it needs to educate people about their own sexual nature, which includes preparation for the pull of polygamy. All the better to resist it, my dear? Perhaps; or not. Either way, all the better to be prepared for those curveballs that our biology is certain to throw our way.

Another downside to underestimating polygamy's gravitational attraction is the prospect—not uncommon—of throwing out the monogamous baby once the bathwater begins to emit odors of polygamy. It is all too easy to assume that once you have defined yourself as monogamous and united with what you imagine to be your life partner, you will never feel polygamous inclinations, much less act on them. And it is equally easy to assume the same for your partner. The problem is that once this naïve expectation is disproven, some people rush to the other side of the aisle, convinced that "I'm not cut out for monogamy"; or that one's partner is similarly flawed.

The truth is that *no one is cut out for monogamy*. For one thing, DNA and our evolutionary heritage isn't a cookie cutter. There is immense interpersonal diversity among human beings, each of whom is "cut out" differently, especially once the vagaries of experience are considered. Moreover, not only is there great diversity among individuals, but each individual is also capable of huge diversity from one moment to the next. People partake of sociosexual arrangements that include celibacy (male or female), polygyny, polyandry, and monogamy—with or without adultery: even, for a small minority, polyamory. The question, therefore, isn't whether someone is monogamy prone, but rather to what extent does he or she *want* to be monogamous (or polygynous, or polyandrous, or celibate, or whatever), and how hard is he or she willing to work to achieve it.

As usual, Darwin got it right when he wrote, in *Sexual Selection and the Descent of Man*, that

> we must acknowledge that man with all his noble qualities, with sympathy which feels for the most debased, with benevolence which extends not only to other men but to the humblest living creature, with his god-like intellect which has penetrated into the movements and constitution of the solar system—with all these exalted powers—Man still bears in his bodily frame the indelible stamp of his lowly origin.

What Darwin didn't say here (but developed at impressive length and persuasiveness in his book, *The Expression of the Emotions in Man and Animals*) is that we also reflect this same stamp—albeit not so "indelible"—of our lowly origin in our behavior, too.

Archaeologists make much of ancient relics—remnants of bone, shards of pottery, flakes of stone—to trace humanity's history. Another treasure trove of ancient *Homo sapiens*, often overlooked but far richer than the usual artifacts that fascinate archaeologists, is the human being of today. Some of these reliquaries are anatomical: vestigial structures such as our appendix or perhaps our tonsils. Some are physiological, embedded in the ebb and flow of various chemical and electrical events. Most interesting for our purposes is behavior, stigmata of our animal past, which even modern, high-tech human beings carry about with them as they rush forward in the 21st century. And prominent in this evolutionary baggage is our long history of polygamy, leavened with monogamy.

Earlier I noted that looking simply at our biology, human beings are rather ordinary mammals. The human claim to specialness is our immense brain power, and the degree to which humanity's behavioral repertoire includes cognition, culture, symbolism, language, and so forth. As a result, we are almost certainly less constrained by our biology than any other species. The eminent British biologist Julian Huxley urged people to avoid the fallacy of "nothing but-ism," the notion that because human beings are animals, they are nothing but animals. His warning was well taken. But it also applies to the alternative, mistakenly assuming that we are nothing but the products of our social learning and cultural traditions.

Anyone wanting to do anything other than "what comes naturally" needs to know what in fact does come naturally, namely, the basic stuff of human nature and sexuality. Some people may then want, specifically, to *avoid* doing what comes naturally. Others might want to celebrate it.

Great harm can be done by misreading the behavioral promptings of our biology, whether by assuming inclinations that aren't there or denying those that exist. Speaking of the perils of the latter, Barbara Kingsolver warned that

> We humans have to grant the presence of some past adaptations, even their unforgivable extremes, if only to admit they are permanent rocks in the stream we're obliged to navigate. A thousand anachronisms dance down the

strands of our DNA from a hidebound tribal post.... If we resent being bound by these ropes, the best hope is to seize them out like snakes, by the throat, look them in the eye and own up to their venom.[11]

Modern evolutionary biology makes it clear that "nature," acting through natural selection, whispers in our ears—cajoling, seducing, imploring, sometimes even threatening or demanding—and undoubtedly inclining us in one direction or another. Whether or not we choose to follow, and if so, how far, is up to us.

Notes

1. Dennett, D. (1996). *Darwin's dangerous idea*. New York: Simon & Schuster.
2. Technically, biologists define a species by the capacity of its members to exchange genes—to interbreed—with each other. In practice, however, species are generally recognized by some aspect of distinctiveness from other individuals that comprise different species.
3. Ghiselin, M. (1974). *The economy of nature and the evolution of sex*. Berkeley: University of California Press.
4. Coontz, S. (2005). *Marriage, a history*. New York: Viking.
5. Foucault, M. (1978). *The history of sexuality, Vol. I*. New York: Random House.
6. Ryan, C., & Jetha, C. (2010). *Sex at dawn*. New York: Harper.
7. Saxon, L. (2012). *Sex at dusk*. New York: CreateSpace.
8. Dixson, A. F. (2009). *Sexual selection and the origins of human mating systems*. Oxford, England: Oxford University Press.
9. Hua, C. (2001). *A society without fathers or husbands: The Na of China*. New York: Zone.
10. Boesch, C. (2002). *Behavioural diversity in chimpanzees and bonobos*. Cambridge, England: Cambridge University Press.
11. Kingsolver, B. (2003). *High tide in Tucson*. New York: Harper.

Afterword

I am delighted to thank my agent, Howard Morhaim, and my editor at Oxford, Jeremy Lewis, for their combination of patience, hard work, and good advice. My debt to my family—notably Judith Eve Lipton—remains both enormous and ever-growing; and I am deeply grateful for the careful research of the many students of evolution, psychology, and anthropology who have provided much of the theoretical structure on which this book has been constructed, as well as the empirical results with which it is clad.

I have been researching and writing about these and related issues for more than four decades, and although most of this book is new, some of it involves repurposing material that I have already written, albeit in different contexts and with different emphasis and focus. Accordingly, here and there, I have reprised some pages—with suitable modifications—that I wrote originally as various articles in the *Chronicle of Higher Education Review* and others that appeared in *Strange Bedfellows: The Surprising Connection Between Sex, Evolution and Monogamy*, coauthored with my wife and originally published by Bellevue Literary Press, as well as several paragraphs originally occurring in *Gender Gap*, published by Transaction, plus others that appeared in my earlier book, *The Hare and the Tortoise* (Penguin). I apologize to any of my earlier readers who—because of my occasional rewriting and ransacking of previous material—find themselves encountering familiar material for the second time. I trust that even in such cases, both the organization and orientation are sufficiently new to warrant a fresh look.

I want to express my appreciation to the students in my University of Washington honors seminar, who read this material in early draft and helped me to clarify the arguments. I have always wanted to write, with regard to one of my books, that the people acknowledged—and not me—are entirely responsible for any errors contained therein . . . but of course, the opposite is true!

Index

Sperm count, polyandry on, 71
Sperm makers, 57
  genetic connection to offspring, 21
  rites of passage, 45–46
  same-sex preference, 22
SRY gene, 45, 46
Stereotypes, sex role, 44–45
Strassman, Beverly, 88–90
Structural powerlessness theory, 92
Suckling, John, "The Constant Lover," 155
Super-weaners, elephant seals, 18–19
Sweeney Todd (Sondheim), 196
Swift, Jonathan, *Gentle Echo on Woman*, 95–96
Symons, Donald, 14
Symposium (Plato), 127–128

Tanner, Nancy, 75–76
Temne (Sierra Leon), 90
Temptation, sexual, 209
Testing of males, female, 77
Testis, 71
Testis size
  human *vs.* other primate, 32
  polyandry, 71
Testosterone, male violence, 44
The African Queen, 198
"The Constant Lover," (Suckling), 155
The King and I, 59
Theodicy, 193
The Origin of Species (Darwin), 202
Timidity, as womanly, 28
Tolstoy, Leo, *Ana Karenina*, 153
Toohey, Peter, *Jealousy*, 149–150
Tre-ba people (Tibet, Nepal), 100, 115
Trial by Jury (Gilbert), 141
Tristan and Isolde (Wagner), 133
Turner's syndrome, 22
Twain, Mark, *Letters From the Earth*, 63
Twain's Devil, 63
23rd Psalm reworked, 174

Unemployment, male violence, 30
Universals, cross-cultural. *See* Cross-cultural universals

Variance, male–female reproductive success, 5, 23n5

harem size, 6
Variety, sexual, male preference, 62–64
  adultery, 143–144
Veiling, Islamic, 188
Verner, Jared, 85–86
Violence, 27–52
  aggressive neglect, 29
  boy–girl differences, 42–43
  dominance hierarchies, 33
  equipotentiality myth, 48–49
  genetics and SRY gene, 45, 46
  hormonal differences and culture, 43–44
  infidelity, 34–37
  male–female, 8, 28
  monogamy *vs.* polygyny, 36–37
  polygynandry, chimpanzees and bonobos, 31–33
  polygyny, 40–41
  poverty, inequality and, 38
  property crimes, 30
  reproductive benefits, 37
  reproductive skew, 38–39
  reproductive success, 38–40
  resource acquisition, 38
  rites of passage, 45–46
  sex role stereotypes, 44–45
  sexual bimaturism, 9–10, 29
  sexual dimorphism, 37
  unemployment, 30
Violence, female
  defensive, 30
  female–female, 51–52
Violence, male, 27
  aggressive, 3–9, 30
  angry young men, 28
  Genghis Khan, 40
  harem masters, 30–31, 37, 41, 50
  homicide, 47–49
  infanticide, 49–51
  Ismail, Moulay, 40
  male–male, 28
  male–male, prison sexual assault, 46
  offspring relations, 33
  pervasiveness, 28
  polygyny, 8–9
  reproductive excess, 40
  sex and, 75–77
  sexual jealousy, 33–34